City & Guilds 3850

Mathematics
FOR CARIBBEAN SCHOOLS

❚ Ann Douglas-Lee

Orders: please contact Bookpoint Ltd, 130 Park Drive, Milton Park, Abingdon, Oxon OX14 4SE. Telephone: +44 (0)1235 827827. Fax: +44 (0)1235 400401. Email education@bookpoint.co.uk Lines are open from 9 a.m. to 5 p.m., Monday to Saturday, with a 24-hour message answering service. You can also order through our website: www.hoddereducation.com

ISBN: 978 1 5104 6327 1

© Hodder & Stoughton Limited 2020
First published in 2020 by
Hodder Education,
An Hachette UK Company
Carmelite House
50 Victoria Embankment
London EC4Y 0DZ

www.hoddereducation.com

Impression number   10 9 8 7 6 5 4 3 2 1

Year   2023 2022 2021 2020

Cover photo © E+/SDI Productions/Getty Images

City & Guilds and the City & Guilds logo are trade marks of The City and Guilds of London Institute. City & Guilds Logo © City & Guilds 2020

Typeset in India.

Printed in India.
A catalogue record for this title is available from the British Library.

Contents

Introduction

It is impossible to go through our daily lives without using mathematics, but the study of the subject can be quite daunting. The City & Guilds Certificate in Mathematics (3850) makes mathematics more accessible and is created for you whether you are a student, a trainee or an adult who desires to improve your mathematical skills either for further education or to take advantage of employment opportunities.

The City and Guilds Certificate in Mathematics (3850) progresses through three levels or stages, the design of which rely on the findings of systematic research into workplace mathematical needs. The course takes a practical approach and by the end you will confidently manoeuvre through practical everyday applications of mathematics.

Stages/levels

Each stage builds on the last, with the aim of steadily improving your understanding of both simple and complex concepts. Each stage provides you with the explanations, examples and activities to increase your confidence in performing mathematical calculations.

By using everyday images and examples such as tables, graphs and pictograms, you see how mathematics forms a basic part of your life and learn how to interpret, compute and use numerical information.

Assessment/examination

Stage	Paper	Format	No. of items	Duration
Stage 1	3850-101	Multiple choice	60	2 hours
Stage 2	3850-102	Multiple choice	60	2 hours
Stage 3	3850-103	Multiple choice	60	2 hours

How to use this book

City & Guilds 3850: Mathematics for Caribbean Schools is structured in three stages with continuous progression of topics and skills demonstrated through everyday practical examples.

Each topic develops from Learning objectives through to Test your knowledge items. To help you move from the introduction of the topic to a demonstration of your mastery are features such as key terms, worked examples, activities and tasks.

- Learning objectives present the expected outcomes for each topic.

- Key terms are the important mathematical terms and concepts. They appear in **blue text** and are explained in the Glossary at the back of the book.

- Worked examples show potential ways of applying mathematical concepts to activities.

- The real world maths feature shows you how these topics apply in the workplace and other aspects of life outside school.

- Activities help you master each segment as you progress through the topic.

- Tasks provide you with an opportunity to demonstrate your understanding of the topic.

- Test your knowledge with practice exam-type questions at the end of each topic.

- Answers for the tasks and Test Your Knowledge questions and selected activities are provided online at www.hoddereducation.com/Caribbean/c-g

Acknowledgements

Every effort has been made to trace all copyright holders, but if any have been inadvertently overlooked, the Publishers will be pleased to make the necessary arrangements at the first opportunity. The Publishers would like to thank the following for permission to reproduce photographs:

Page 22 © Auremar/stock.adobe.com; p.27 © Georgerudy/stock.adobe.com; p.30 © YKD/stock.adobe.com; p.39 © JGI/Getty Images; p.51 © Danny Hooks/stock.adobe.com; p.56 © Hugh O'Neill/stock.adobe.com; p.72 © Vidady/stock.adobe.com; p.82 © Monkey Business Images/Shutterstock.com; p.104 © Asafeliason/stock.adobe.com; p.106 © Flairimages/stock.adobe.com; p.112 © WavebreakmediaMicro/stock.adobe.com; p.118 © Dario Lo Presti/stock.adobe.com; p.119 © Maksym Yemelyanov – Fotolia; p.121 © Daisy Daisy/stock.adobe.com; p.123 © GW3NDOL!N/stock.adobe.com; p.132 © WavebreakMediaMicro/stock.adobe.com; p.133 © Wayhome Studio/stock.adobe.com; p.141 © Andrey Popov/stock.adobe.com; p.150 © Paulo/stock.adobe.com; p.157 © Mimagephotos/stock.adobe.com; p.164 © mavoimages/stock.adobe.com; p.167 © LIGHTFIELD STUDIOS/stock.adobe.com; p.175 © Akepong/stock.adobe.com; p.178 © O.Farion/stock.adobe.com; p.186 © Tetra Images, LLC/Don Mason/Alamy Stock Photo; p.188 © Simbos/stock.adobe.com; p.195 © Aliaksei/stock.adobe.com; p.200 © Rawpixel.com/stock.adobe.com; p.202 © Stephen Coburn – Fotolia.com; p.212 © Kablonk RM/Golden Pixels LLC/Alamy Stock Photo; p.233 © Leonid/stock.adobe.com; p.252 © LVL/stock.adobe.com; p.258 © Fotofabrika/stock.adobe.com; p.261 © Anne/stock.adobe.com; p.262 © 169169/stock.adobe.com; p.265 © Ariadne Van Zandbergen/Alamy Stock Photo; p.272 © Remus20/stock.adobe.com; p.273 © Ngrigoryeva/stock.adobe.com; p.278 © Fotoatelie/stock.adobe.com; p.283 © Jeremy Graham/Dbimages/Alamy Stock Photo; p.293 © Joraduet/stock.adobe.com; p.302 © Alexandr_DG/stock.adobe.com; p.306 © Mariusz Blach/stock.adobe.com; p.308 © Stockbyte/Entertainment & Leisure CD35/Getty Images; p.312 © Rawpixel.com/Shutterstock.com; p.316 © Andrey gonchar/stock.adobe.com; p.319 © Rabbit75_fot/stock.adobe.com; p.323 © Torriphoto/stock.adobe.com; p.336 © Nabil BIYAHMADINE/Fotolia.com; p.347 © Nickolae – Fotolia.com; p.349 © Panya168/stock.adobe.com; p.360 © Sergey Novikov/stock.adobe.com; p.370 © Icholakov/stock.adobe.com.

Unit 101
Numbers and the number system

Introduction

We use numbers all the time in our everyday lives for counting money, putting things in size order or describing parts of amounts. Sometimes we need to be exact, and sometimes a close amount will do.

At Stage 1, the emphasis is on developing the basic number skills of counting, ordering numbers up to thousands, and developing early ideas about **estimating** and **rounding**. Much of the rest of mathematics at this level is built on this.

Learning objectives

In this unit, you will find information on:

● numbers up to thousands, including number patterns and simple fractions

● estimating and rounding numbers.

This will help you to prepare for questions about:

● using place values to read 3-digit numbers accurately

● completing a number pattern.

Place value

Numbers are everywhere. Sometimes they are in order and sometimes we need to put them in order to find out what we need to know. For example, we may need to find the highest score or whether we have enough money.

You should know and understand which **digit** of a number shows the number of hundreds, which shows the number of tens and which shows the number of units – this is known as **place value**. This section helps you to check that you know enough about the digits of a number before you build more skills at this level.

56

This number reads fifty-six.

It means five tens and six units.

If you carry on counting, you eventually get to numbers with three digits. Look at this tape measure, which shows the movement from two digits to three digits:

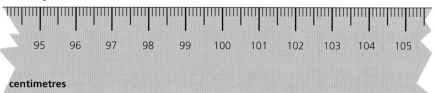

| A tape measure is a type of number line. |

centimetres

You can think about the way numbers grow by building them up. Look at the grid with different hundreds, tens and units.

Hundreds	Tens	Units
900	90	9
800	80	8
700	70	7
600	60	6
500	50	5
400	40	4
300	30	3
200	20	2
100	10	1

So the number **five hundred and thirty-seven** is made by five hundreds (**500**), three tens (**530**), and seven units (**537**).

Think what happens with the number three hundred and eight.

This is made up of three hundreds (300), and eight units (308).

| Notice we don't have any tens so the zero stays in the tens position and 8 goes in the units column. |

What would the number be if we put 8 in the tens column and 0 in the units column? This would be 380.

> **Activity**
>
> **1** In pairs, take turns to create and say a 3-digit number for your partner to break down and write in numbers.
>
> **2** In pairs, take turns to create and write down a 3-digit number for your partner to break down and write in words.

Counting on

The grid shows how you can count up in tens or hundreds.

So counting on in tens, starting at **40** is fifty, sixty, seventy, eighty, ninety but then we are at the top of that column, so you have to say one hundred. Still counting on in tens gives one hundred and ten, one hundred and twenty, and so on.

Hundreds	Tens	Units
900	90	9
800	80	8
700	70	7
600	60	6
500	50	5
400	40	4
300	30	3
200	20	2
100	10	1

In your own head, but looking at the grid, carry on counting up in tens until you get to three hundred and forty. Imagine circles appearing and disappearing round the numbers as you count. Three hundred and forty is made by three hundreds (300), four tens (340), and no units (340).

Activity

In pairs, take turns to give your partner two 3-digit numbers. Your partner should count on in tens from the lower number to the higher number. Remember, both of the numbers should end in the same digit from the units column for this activity.

Ordering and comparing numbers

The grid helps in understanding how to order numbers, but you're also going to think about numbers on a number line, like the tape measure.

Compare the numbers **two hundred and thirty-five** with three hundred and twenty-seven. Starting with the largest part of the number, the hundreds, 300 is more than **200**, so three hundred and twenty-seven (327) is larger than two hundred and thirty-five (**235**).

Hundreds	Tens	Units
900	90	9
800	80	8
700	70	7
600	60	6
500	50	5
400	40	4
300	**30**	3
200	20	2
100	10	1

Real world maths

If you are able to, use Jamaican $100 notes, $10 and $1 coins to count to 340.

Real world maths

Here is another way to think about it: if you owned a beachwear boutique and needed to purchase three hundred and twenty-seven pairs of flip-flops for the summer, would you go to a supplier who has no more than two hundred and thirty-five pairs of flip-flops for sale?

Now what about five hundred and twenty-nine compared with five hundred and ninety-one? The largest part of the number in both cases is five hundred, so we need to look at the tens. As ninety is bigger than twenty, this makes the five hundred and ninety-one (591) bigger than the five hundred and twenty-nine (529).

You can show this on a number line, which is like a tape measure.

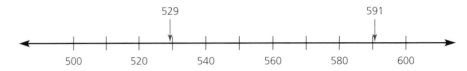

Number patterns

You have probably heard of **odd** numbers and **even** numbers.

Odd numbers are not evenly divisible by 2 and leave a remainder of 1. So 1, 3, 5, 7, 9, etc. are odd numbers because if we divide them by 2 we have a remainder of 1 left.

Even numbers are evenly divisible by 2 without leaving a remainder. So 2, 4, 6, 8, 10, etc, are even numbers. We can divide any of them by 2 and have no remainder left.

Odd and even numbers do not stop at 10: larger numbers can be described as odd or even.

What about 41 and 18? 41 is an odd number because if we divide 41 by 2 we have a remainder of 1 left. 18 is an even number because it is evenly divisible by 2.

Real world maths

Which of the sports listed below is made up of an even number of players on each team?

a volleyball

b cricket

What is the pattern in this sequence?	Answer	Explanation
2, 4, 6 . . .	These are all even numbers so the next number is 8.	Each number is 2 more than the previous number.
3, 6, 9 . . .	The next number is 12.	Each number is 3 more than the previous number.
5, 10, 15 . . .	The next number is 20.	Each number is 5 more than the previous number.

Number patterns also include square numbers.

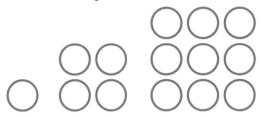

The next pattern in the sequence would be:

In pairs take it in turns to make a number pattern and ask your partner the next number in the sequence.

Using a symbol to stand for an unknown number

Often we want to know the answer to an addition sum. For example:

$3 + 6 = 9$

We could ask $3 + ? = 9$ or $3 + Y = 9$

Note that ? and Y are both used to stand for the same missing number.

The missing number would be 6, so $? = 6$ and $Y = 6$ in these examples.

$2 \times \lozenge = 6$

The missing number is 3, so $\lozenge = 3$.

Activity

In pairs take it in turns to make a simple sum with a missing number and ask your partner to find the missing number in the sum.

Introduction to decimals and simple fractions

Mathematics has three ways to describe parts of numbers: using a **fraction**, a **decimal** or a **percentage**. This unit will introduce fractions and decimals and look at how to notice when they show the same amount, but in two different ways.

Decimal numbers

Remember the Hundreds, Tens and Units grid we looked at earlier. Each unit is made up of smaller parts called tenths and hundredths.

Hundreds	Tens	Units	.	Tenths	Hundredths
900	90	9		9	90
800	80	8		8	80
700	70	7		7	70
600	60	6		6	60
500	50	5		5	50
400	40	4		4	40
300	30	3		3	30
200	20	2		2	20
100	10	1		1	10

The first number after the decimal point represents tenths, the next represents hundredths, and so on. You will be familiar with this from using money. The dollar is made up of 100 cents, so half a dollar is 50 cents. Using the grid, this is 5 tenths of a dollar and 0 hundredths.

We can write this as $0.50

Simple fractions

A fraction is a way of describing a number that is less than one whole. Think of a pizza.

Here it is sliced into two pieces or **halves**. As a fraction of the whole pizza, we write that one slice is a $\frac{1}{2}$.

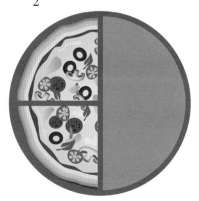

This is said 'one-half'.

In a fraction, the top number is called the **numerator** and tells you how many parts the fraction shows. The bottom number is called the **denominator** and tells you how many parts the whole is split into.

So if the pizza is split into four pieces, one piece is $\frac{1}{4}$.

This is said as 'one-quarter'.

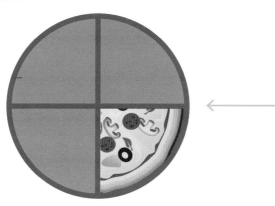

Notice the size of the slice for one quarter of a pizza is smaller than for one half. You can see that two of these slices would give you the same amount of pizza to eat as having one half of the first pizza.

Activity

1 Take a piece of paper and fold it into 2 equal halves. Then cut along this fold line.

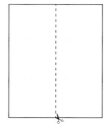

2 Next fold each of the two pieces of paper into 2 equal halves and cut along the fold line. You can fold in one of two ways so your pieces of paper may look like this

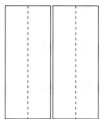

Or like this

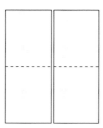

What would the fraction $\frac{3}{4}$ look like as a pizza picture? This would be a pizza cut into four slices (the bottom number or denominator) and you taking three of these slices (the top number or numerator) to eat.

Activity

Use your pieces of paper to make the page whole again and then take three-quarters away. What fraction is left?

Real world maths

Fractions don't just apply to slices of a whole. If you have 6 biscuits and eat one-half of them, how many have you eaten? This is the same as making two equal piles of biscuits (remember the bottom number, or denominator, is dividing into two here). The top number tells us how many parts or piles, so here it is one pile of three biscuits.

$\frac{1}{2}$ of 6 = 3

If you have 10 sweets and give your friend half of them, how many would he have? Make two equal piles (remember the bottom number or denominator is dividing into two here) and give your friend half.

$\frac{1}{2}$ of 10 = 5

If you have 8 biscuits and eat one quarter of them, how many have you eaten? This is the same as making four equal piles of biscuits (remember the bottom number, or denominator, is dividing into four here). The top number tells us how many parts or piles, so here it is one pile out of four in total. Each pile has two biscuits.

$\frac{1}{4}$ of 8 = 2

Activity

Try finding half of different even numbers. You can use counters, pencils or cents to help you.

What happens if you have an odd number, like 5? You can make two equal piles of 2 but you have one left over. If this is a biscuit you could break this in half and then have two equal piles with $2\frac{1}{2}$ biscuits in each pile. Half of 5 is $2\frac{1}{2}$.

If you want to find half of a pile of 7 sweets, you can make 2 piles of 3 with one left over.

Half of 7 is $3\frac{1}{2}$.

Tips for assessment

Know and quickly recognise what different parts of numbers mean, and their place value. Having the image of the grid and the number line in your head, so that you can sketch it, is really helpful until you get confident about using numbers.

Know that the denominator is the bottom of a fraction and shows how many parts the whole has been split into. Have an image in your head of a pizza cut into parts, so you can compare the sizes of different fractions.

Equivalencies

We also use fractions in measures, weights and capacity. A metre is a common measurement. You may have seen one-half of a metre written as 0.5 m or one-quarter of a kilogram written as 0.25 kg.

Some important equivalent values to remember are given in the table below.

Image of fraction	Fraction	Decimal
	$\frac{1}{2}$	0.5
	$\frac{1}{4}$	0.25
	$\frac{1}{10}$	0.1

Rounding and approximating

Sometimes you only need to use an approximate number, such as when counting large groups of people or making sure you have enough money to pay a bill. When you are paying for an item, you may not have the exact notes and coins you need, so you offer a convenient overpayment and the shop assistant will give you the required change.

An item costs $17. You have $10 and $20 bills. 17 is between 10 and 20, so you know that $10 will not be enough to pay for the item. You pay with a $20 bill and receive change. This is known as **rounding**.

You will find out more about rounding in Unit 208.

> ### Activity
> In pairs, take it in turns to ask each other questions involving rounding.

Tasks

Now have a go at these tasks.

1 a Copy the table below and for each number, circle the hundreds, tens and units that make the number, on the grid. Use a different colour for each number.

900	90	9
800	80	8
700	70	7
600	60	6
500	50	5
400	40	4
300	30	3
200	20	2
100	10	1

   i   652
   ii   419
   iii 340
   iv   68
   v   905
   vi   276

   b Now put these numbers in order, smallest to largest.

2 a Copy and complete the number line below so that it goes up to 400. 200 has already been marked for you.

   200

   b Use the number line to mark, with an arrow, the position of each of the numbers below. Remember to label each arrow.
      Sometimes the number may be between graduation marks on the line.

      230   320   315   395   218   279   195   342

3 What is the next number in the sequence?
   3, 6, 9, 12, ?

4 Draw the pattern for the next triangular number.

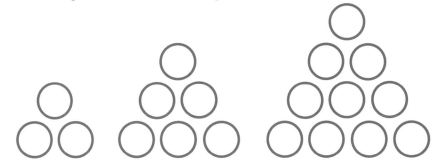

5 What is the missing number?

   a $10 - \diamond = 4$

   b $4 \times ? = 12$

   c $\wedge + 2 = 7$

6 a Write 13 408 in words.

   b Write seventeen thousand and six in figures.

7 Write down the fraction of pizza that is left in these pizza pictures.

   a   **b**

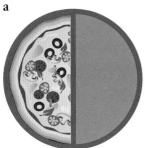 

8 Match the fraction with the shaded portion in these pictures.

   $\dfrac{1}{4}$ $\dfrac{1}{2}$ $\dfrac{3}{4}$

9 a What is one-half of eight?

   b What is one-quarter of eight?

10 a What is $\dfrac{1}{2}$ of 12?

   b What is $\dfrac{1}{4}$ of 12?

11 Match the fraction with the decimal fraction.

   0.5 $\dfrac{3}{4}$

   0.75 $\dfrac{1}{4}$

   0.25 $\dfrac{1}{2}$

12 What is the smallest coin that can be used to pay for an item costing 44¢?

In your exam, the questions are multiple choice. This means that you need to choose the correct answer from the four options given: a, b, c and d. The correct answer will always be shown, but the other answers often look reasonable and may be the result of a simple mistake or an error in the method.

For example, which of these numbers has 6 hundreds?

    a 286

    b 682

    c 268

    d 862

All the options have a 6 in them, but only one has 6 in the hundreds column so (b) is the correct answer.

Try these

1 Which is the third highest number?

280   476   712   689

    a 280

    b 476

    c 712

    d 689

2 Put these numbers in order, smallest to largest.

235   367   280   189   198   356

    a 189, 235, 198, 280, 356, 367

    b 280, 235, 356, 367, 198, 189

    c 367, 356, 280, 235, 198, 189

    d 189, 198, 235, 280, 356, 367

3 Which of these numbers has 2 tens?

    a 286

    b 628

    c 268

    d 862

4 What is the next number in the sequence?

10, 20, 30, ?

    a 31

    b 40

    c 50

    d 60

5 7 − X = 4

So X =

  a 3

  b 4

  c 7

  d 11

6 What is ten thousand and fourteen in figures?

  a 1 014

  b 10 014

  c 100 014

  d 1 000 014

7 What fraction of the chocolate bar is shaded?

  a $\frac{2}{6}$

  b $\frac{1}{8}$

  c $\frac{1}{2}$

  d $\frac{1}{4}$

8 What is 25 cents written in dollars?

  a $25

  b $2.5

  c $0.25

  d $0.025

9 What is $\frac{1}{2}$ kg written as a decimal number?

  a 1.2

  b 0.12

  c 0.2

  d 0.5

10 What is the smallest dollar bill that can be used to pay for an item costing $41?

  a $1

  b $20

  c $50

  d $100

Introduction

This unit is about using measure and time to solve everyday problems.

Length	measuring how long things are, or how far apart things are
Time	finding what time of day it is
Weight/mass	measuring how heavy things are
Capacity	measuring amounts of liquid
Temperature	measure how hot things are

Learning objectives

In the unit, you will find information on:

- estimating and measuring
- the units to use when recording and talking about measures including relating different units within a system
- the tools we use to measure
- temperature facts
- reading and understanding measuring tools
- following and giving directions and instructions including rotation and the four main compass points.

This will help you prepare for questions about:

- reading from a scale to find a weight, length, capacity or temperature and comparing with a similar measure
- stating where one item, or place, is in relation to another
- following and giving directions
- reading a clock time and converting between the 12-hour clock and the 24-hour clock.

Measuring and estimating length

Length is a measure of how long things are.

We also use these words

- **distance** = how far apart things are
- **width** = how wide things are
- **height** = how tall or high things are

Real world maths

We need to measure length accurately in many situations:

- knowing that items will fit in a space, for example fitting a bed into a room
- cutting materials to make things, for example a piece of wood to make a shelf
- knowing how far to travel to get to a place.

Common units for the distance between places are:

miles	kilometres
An average person could walk one mile in 20 to 25 minutes.	An average person could walk one kilometre in 10 to 15 minutes.
A car could travel one mile in a minute on a fast road.	A car could travel one kilometre in under a minute on a fast road.

Common units for length are:

millimetres	centimetres	metres
A millimetre is a very small length. It is roughly the thickness of a coin.	A centimetre is roughly the width of a little finger.	A metre is roughly the length of one stride.
A millimetre is often written as mm, so 30 mm is short for 30 millimetres.	A centimetre is often written as cm, so 25 cm is short for 25 centimetres.	A metre is often written as m, so 5 m is short for 5 metres.
We often use millimetres to measure very short lengths, such as a 12 mm screw.	We use centimetres to measure everyday items like furniture.	We use metres to measure longer lengths such as a room or a running race.
Builders and engineers also measure longer lengths in millimetres to be very precise.	A centimetre is the same as 10 millimetres	One metre is the same as 100 centimetres.
There are 10 millimetres in 1 centimetre. There are 1000 millimetres in 1 metre.	There are 100 centimetres in 1 metre	One metre is also the same as 1000 millimetres.

Here are some common measurements.

Item	Measurement
Width of a bank card	85 mm
Length of a sheet of letter size paper	Roughly 28 cm
Height of an average adult	About 180 cm
Length of a small car	About 4 m
Length of a bus	About 8 m
Length of a football pitch	100 m

Estimating length and distance

Sometimes we do not need an accurate measurement and can say for example that our school is about 2 km from our house. This tells another person that we do not live close to the school but we can walk the distance. The actual distance might be under 2 km but it is more than 1 km. This is an **estimate** or approximation.

Using scales

We measure lengths using a ruler or tape measure.

Measuring lengths means reading the scale.

This diagram shows a ruler marked in millimetres.

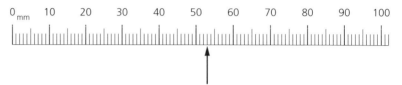

The pointer is between 50 mm and 60 mm.

It is pointing at the 3rd division after 50 mm, which is 53 mm.

Some rulers are marked in centimetres, like this one:

> On this ruler, the small divisions are millimetres, and the large ones are centimetres.

The pointer is between 5 cm and 6 cm.

It is pointing at the 3rd division after 5 cm, which is 5 cm 3 mm, or 53 mm.

Example 1

Grace needs a label of length between 45 mm and 55 mm.

She measures the label in this diagram.

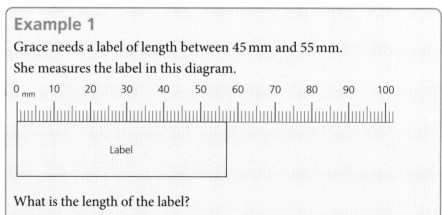

What is the length of the label?

Make one comment about whether Grace should use the label.

Solution

The label is between 50 mm and 60 mm long.

It comes up to the 7th division after 50 mm, so it is 57 mm long.

Grace wants a label between 45 mm and 55 mm, so it is too long.

Example 2

Daryl wants to measure his height.

The pointer on this scale shows his height.

Daryl thinks his height is 164 cm.

What is his height? Is he correct?

Solution

The pointer is between 160 cm and 170 cm.

There are 10 divisions between 160 cm and 170 cm.

Those 10 divisions cover 10 cm, so 1 division is 1 cm.

The pointer is at the 6th division after 160 cm, so the height is 166 cm.

Daryl is wrong. He is taller than he thinks.

Real world maths

We use measurements every day for different purposes. For example, when we buy fabric to make clothes or materials to build a house.

Activity

Copy and complete the table below.

Estimate the length first and then use a ruler or tape measure to find the actual length. Choose a suitable unit of measurement.

Item	Estimate	Actual measurement
Width of a table		
Height of a door		
The measurement around your wrist		
The length of your pencil or pen		

Don't forget to say which units you have used for each measurement.

Tasks

Now have a go at these tasks.

1 What is the length shown by the pointer?

2 What is the length shown by the pointer?

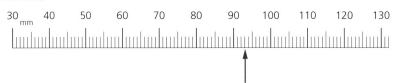

3 What is the length shown by the pointer? Choose option a, b, c or d from the list.

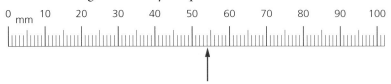

a 90 mm

b 93 mm

c 95 mm

d 107 mm

4 Juliet wants to buy a TV stand.

She measures the length of the space for a TV stand in her room.

The pointer shows the length.

a What is the length of the space for the TV stand shown on the measuring tape?

Juliet finds this information on TV stands in a shop.

TV stand	Length (cm)
A	65
B	72
C	75
D	78

b Which is the longest TV stand that will fit in this space?

5 Robert is at a theme park.

   a He measures his height to see which rides he can go on.
      The pointer on this scale shows his height.

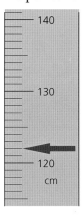

      What is Robert's height?

   b These are the rides.

Ride	Minimum height (cm)
Swirler	95
Splash	115
Spooks	125
Big swing	135
Terror-ride	145

      Which rides can Robert go on?

Following and giving directions and instructions

Compass points

As well as distance, we often want to know where one item, or place, is in relation to another. For example, in which direction should we walk or drive?

This shows the four **points of the compass**: North (N), East (E), South (S) and West (W). A compass is an instrument that shows magnetic North. Therefore, if your house faces North it will always face North. The sun will rise and pass over your house but North will always be in the same place. A compass is used for navigation, location and direction. People often use a compass to find their way on a hiking trip or when travelling to a new location.

Activity

In pairs, take turns to ask your partner a question related to the compass.

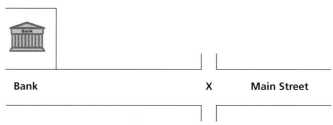

| Bank | X | Main Street |

We need to go West.

If we are on Main Street at point X and want to go to the bank, which way should we go? Think of the compass.

Rotation

Rotation is the process or act of turning or circling around something. An example of rotation is the earth's orbit around the sun. A person can start facing North and turn a complete circle (a complete rotation) and they will be facing North again. If they turn halfway this is called a half rotation and, in this example, they would be facing South.

Real world maths

You may have to turn a tap or screw a half or a quarter rotation, or more, to turn the tap on or tighten the screw. You would turn the tap or screw one way to tighten and the other way to loosen so it is important to know which way to turn.

Clockwise rotation

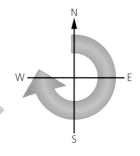

Turning from North to East is a one-quarter rotation as this is a quarter of the whole (see page 7). Turning from North to South is a one-half rotation. What is turning from North to West? ←

$\frac{3}{4}$ rotation

It is important to know which way to turn. You would turn a tap or screw one way to tighten and the other way to loosen. **Clockwise**, as in the diagram above, is in the direction of a clock's hands. **Anticlockwise** is the opposite way. You don't have to start at the top, as this diagram shows.

Activity

Face North. In pairs, take turns to give your partner an instruction regarding rotation (clockwise or anticlockwise). Which point of the compass is your partner facing now?

clockwise   anticlockwise

Left and right

You should also know which hand is your left hand and which is your right hand. Some people remember this by the hand they write with. A lot of people write with their right hand and these words sound the same.

Another way of remembering which is left and right is to put your hands flat on the table in front of you.

The hand that forms an 'L' is your left hand and your left side.

When giving directions, people often say something like, 'Go to the end of the road and turn left (or right)'.

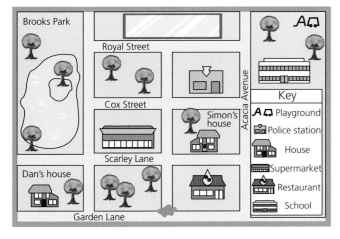

How do you walk from the school to the restaurant? ←

It can be helpful to imagine yourself walking down the road. Some people find it easier to turn the map round.

Turn left down Acacia Avenue and the restaurant is at the bottom on your right.

> ### Activity
> Use the simple map on page 21 or a local map. Tell your partner where to start and give him or her directions. Where are they now?
>
> Take it in turns to give directions and follow directions.

Tasks

Now have a go at these tasks.

1 You are facing North and turn a quarter turn anticlockwise. Where are you facing now?

2 Explain how to walk between your classroom and another place at your school. Use left and/or right.

3 You are in the car park on West Drive. Turn left and take the first right into Central Avenue.

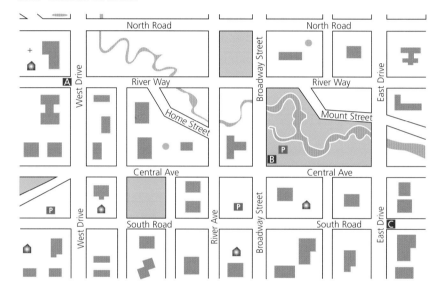

a Take 2nd left and then first left. Where are you?

b Give instructions to get from A to C.

c Give instructions to get from C to B.

Time

Measuring time is part of everyday life at home, work and leisure.

This includes

- using a calendar or diary
- understanding dates expressed in different ways
- reading clocks
- understanding times expressed in different ways.

Calendars and dates

We use a calendar or diary to keep track of important dates for work, holidays, appointments and to plan meetings and events.

This is a common layout for a calendar.

October 2019

Monday	Tuesday	Wednesday	Thursday	Friday	Saturday	Sunday
	1	2	3	4	5	6
7	8	9	10	11	12	13
14	15 Holiday	16 Holiday	17 Holiday	18	19	20
21	22	23	24	25	26	27
28	29	30	31			

The dates are shown by the numbers.

There is space for each day for writing in important events.

This calendar can be used to find answers to questions such as:

Which day is the 22nd October? ←

Answer: Tuesday (22 is in the Tuesday column.)

What date is the last Sunday in October? ←

Answer: The 27th (The last number in the column for Sunday is 27.)

Example

You buy something online on the 11th and it will take 5 days to be delivered. Will you be at work or on holiday when it is delivered?

Solution

Counting 5 days on from the 11th is the 16th. This is a holiday.

Real world maths

The calendar can also be used to plan events.

Activity

In pairs, take turns to ask and answer questions using a calendar.

Writing dates

Dates can be written in different ways.

The 21st October 2019 can be written in figures:

21/10/2019

This is the day – the 21st

This is the month. The
10th month is October.

This is the year.

There is a shorter form with just the last two digits of the year: 21/10/19

The numbers are always in the same order, day/month/year.

On a computer, the date is sometimes written like this: 21/Oct/2019

These are the numbers and shortened words for the months:

January	Jan	01
February	Feb	02
March	Mar	03
April	Apr	04
May	May	05
June	Jun	06
July	Jul	07
August	Aug	08
September	Sep	09
October	Oct	10
November	Nov	11
December	Dec	12

Activity

In pairs, take turns to ask questions and write down dates such as birthdays and special community or school events.

Note that the month numbers 1 to 9 have a zero in front of them.

This is because many websites expect a 2-digit number for the month.

How would you write 11th January 2020?

This could be 11 Jan 20 or 11/01/20.

Time on clocks

Some clocks have digital displays like this:

08:05

The two digits on the left show the hour.

The two digits on the right show the minutes past the hour.

This time is expressed as:
eight 'o' five
five minutes past eight
five past eight.

This table shows how we read other times.

08:10	**Ten past eight**
08:15	**Quarter past eight.** There are 60 minutes in an hour, so 15 minutes is a quarter of an hour.
08:20	**Twenty past eight**
08:25	**Twenty-five past eight**
08:30	**Half past eight.** There are 60 minutes in an hour, so 30 minutes is half an hour.
08:35	**Twenty-five to nine.** When it is 35 minutes past 8, it is the same as 25 minutes before 9.
08:40	**Twenty to nine.** When it is 40 minutes past 8, it is the same as 20 minutes before 9.
08:45	**Quarter to nine.** When it is 45 minutes past 8, it is the same as 15 minutes (a quarter of an hour) before 9.
08:50	**Ten to nine.** When it is 50 minutes past 8, it is the same as 10 minutes before 9.
08:55	**Five to nine.** When it is 55 minutes past 8, it is the same as 5 minutes before 9.
09:00	**Nine o'clock**

An analogue clock uses pointers called **hands** to show the time.

The shorter pointer is the **hour hand**.

It points to the hour at the o'clock time – at 8 o'clock, it points straight at 8.

As the time gets nearer to 9 o'clock, the hour hand moves towards the 9.

At half past eight, it is half way between 8 and 9.

The longer pointer is the **minute hand**.

It points straight up at the o'clock time.

As time goes on it moves round and shows the minutes past the hour.

There are **60** minutes in an hour. At 'half-past' the longer pointer or minute hand has moved to the bottom of the clock to number 6. If you count up the minutes between 'o'clock' and 'half past' you will see there are 30 minutes. 30 is half of 60 (see fractions section in Unit 101, page 6). We say eight-thirty or half-past but not 30 minutes past.

We say **past** for minutes up to half past then **to** for minutes between half past and o'clock.

Three-quarters of the way round the clock is 45 minutes and we say 'eight forty-five' or **'quarter to'**.

15 is one-quarter of 60. When the minute hand has moved 15 minutes from 'o' clock' we say eight-fifteen or '**a quarter past**'.

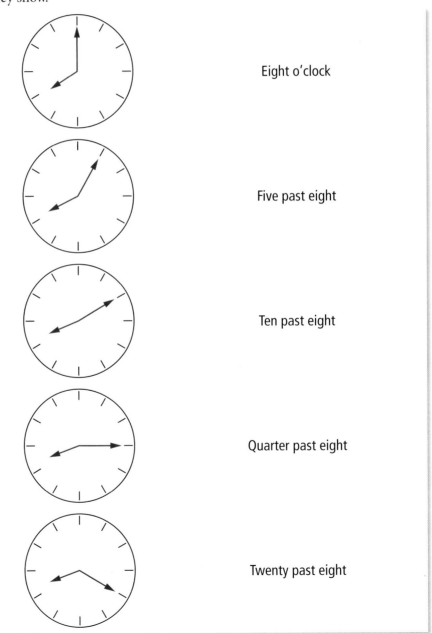

This table shows some different positions of the minute hand and the times they show.

Eight o'clock

Five past eight

Ten past eight

Quarter past eight

Twenty past eight

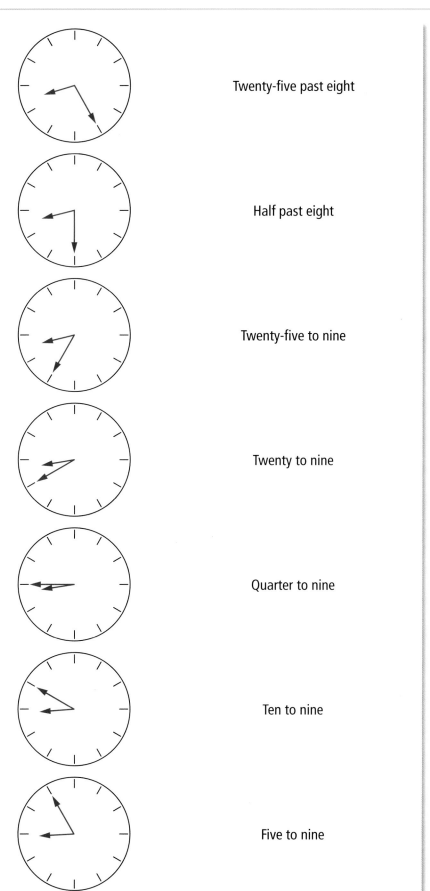

Twenty-five past eight

Half past eight

Twenty-five to nine

Twenty to nine

Quarter to nine

Ten to nine

Five to nine

> It is important to know which!
>
> Your holiday flight leaves at twenty past eight. Do you go to the airport in the morning or the evening? If you get it wrong, you could be much too early or miss the flight.

Morning or afternoon?

This digital clock shows twenty past eight.

08:20

It could be twenty past eight in the morning or twenty past eight in the evening.

There are two ways of showing morning or afternoon: a.m. and p.m.

A morning time is shown by the letters a.m. after the time.

8.20 a.m. is twenty past eight in the morning.

An afternoon or evening time is shown by the letters p.m. after the time.

8.20 p.m. is twenty past eight in the evening.

Midnight is 00.00 a.m.

Midday is 12.00 p.m.

24-hour clock

The 24-hour clock is used in timetables.

In the 24-hour clock, times are written as 4 digits, sometimes separated by a colon (:)

The numbers for the hours go from 00 up to 23.

This table shows the same times in 12-hour and 24-hour clock.

Morning		Afternoon	
12-hour clock time	24-hour clock time	12-hour clock time	24-hour clock time
0.00 a.m.	00:00	12.00 p.m.	12:00
1.00 a.m.	01:00	1.00 p.m.	13:00
2.00 a.m.	02:00	2.00 p.m.	14:00
3.00 a.m.	03:00	3.00 p.m.	15:00
4.00 a.m.	04:00	4.00 p.m.	16:00
5.00 a.m.	05:00	5.00 p.m.	17:00
6.00 a.m.	06:00	6.00 p.m.	18:00
7.00 a.m.	07:00	7.00 p.m.	19:00
8.00 a.m.	08:00	8.00 p.m.	20:00
9.00 a.m.	09:00	9.00 p.m.	21:00
10.00 a.m.	10:00	10.00 p.m.	22:00
11.00 a.m.	11:00	11.00 p.m.	23:00

Midnight is 00:00

Midday is 12:00

Reading from the table, 20:00 in 24-hour clock is the same as 8.00 p.m.

To write 4.00 p.m. in 24-hour time, use 16:00.

The morning times use the same digits in 12-hour and 24-hour time.

The afternoon times can be tricky. ←

An easy mistake is to think 19:00 is 9.00 p.m., but that is wrong. 19:00 is 7.00 p.m.

> This is a quick way to change an afternoon 24-hour clock time to a 12-hour clock time: take 12 away from the hour in the 24-clock time.
>
> Don't forget the p.m.

Example

What time is 19:45?

Solution

The hour is 19. 19 − 12 = 7

19:45 is the same as 7.45 p.m.

Activity

Try changing between 24-hour clock times and 12-hour clock times.

Complete the table. The first one is done for you.

24-hour clock	12-hour clock
09:00	9.00 a.m.
15:00	
	4.20 a.m.
20:30	
23:45	
	10.25 p.m.
	5.50 p.m.
13:05	
18:35	
	7.10 p.m.
	2.10 a.m.

> **Tips for assessment**
>
> Write dates in figures like this: day number/month number/year number
>
> Remember a correct example, such as your birthday: 21/02/98 or 21/02/1998.
>
> Write times in 24-hour clock, or in 12-hour clock with a.m. or p.m.
>
> For example: 15:30 or 3.30 p.m.
>
> Remember to add a.m. or p.m. on 12-hour clock times!

Tasks

1 How else could you write 3rd April 2020?

2 How else could you write 01/06/22?

3 What time does this clock show?

4 Write the time shown in Question 3 as a 24-hour clock time for morning and evening.

5 Your bus should arrive at 3.15 p.m. It arrives at 20 minutes to 4. How many minutes late is the bus?

Measuring and estimating weight

Weight is a measure of how heavy things are.

We need to measure weights accurately in many situations:

- measuring the right amounts of ingredients for cooking
- making sure bags are not too heavy at the airport
- paying the right amount to send a parcel at the post office.

Common units for weight are:

> We use grams for lighter weights and kilograms for heavier weights.

grams	kilograms
A gram is a very small weight. A peanut weighs one gram.	A kilogram is a heavier weight. A bag of sugar weighs one kilogram.
A gram is often written as g, so 50 g is short for 50 grams.	A kilogram is often written as kg, so 25 kg is short for 25 kilograms.

There are 1000 grams in a kilogram.

Here are the weights of some common items.

Item	Weight
Sheet of letter size paper	4.5 g
Bag of chips	30 g
Bag of sweets	200 g
Tin of beans	400 g
Bag of sugar	1 kg
Chicken	2 kg
Baby	3 kg
Suitcase	20 kg
Adult man	90 kg
Car	1200 kg

The pointer shows the weight of the item.

You can see that the pointer is between 100 grams and 200 grams.

To work out the exact weight, we need to know what weight the divisions show (shown by the small lines).

Using weighing scales

Using weighing instruments often means reading a scale.

This diagram shows the sort of scale found on weighing machines.

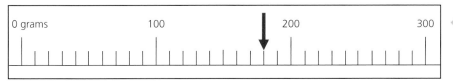

You can check by counting 10s for each division. The first division after 100 grams is 110 grams, then the next is 120 grams and so on up to 200.

There are 10 divisions between 100 grams and 200 grams.

Those 10 divisions cover 100 grams.

So, each division is 10 grams.

There are 8 divisions between 100 grams and the pointer.

8 divisions are 80 grams.

80 grams more than 100 is **180 grams** – that is the weight shown.

On different weighing machines, the unlabelled divisions might not be 10 grams.

They could be 10 grams, 100 grams, 1 kg, 10 kg or other divisions.

To check, there are 2 divisions between the pointer and 200 grams.

2 divisions are 20 grams.

20 grams less than 200 grams is **180 grams.**

Look at each small division on these scales. How much does each small division on these scales represent?

1 kg →

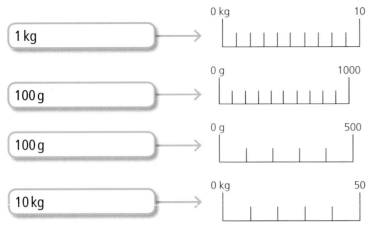

100 g →

100 g →

10 kg →

Example

Maya is going on holiday.

The pointer on this scale shows the weight of her suitcase.

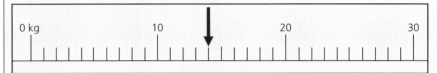

She is allowed to take a maximum of 18 kg of luggage.

How much more weight can she add to her suitcase?

Solution

The pointer shows a weight between 10 kg and 20 kg.

There are 10 small gaps between 10 kg and 20 kg, so each small gap is 1 kg.

There are 4 gaps from 10 kg to the pointer.

The weight is 10 + 4 = 14 kg.

She is allowed 18 kg.

18 − 14 = 4

She can add 4 kg to her suitcase.

Activity

Copy and complete the table below.

Estimate the weight and then use a weighing scale to find the actual weight. Choose a suitable unit of measurement.

Don't forget to say which units you have used for each measurement! →

Item	Estimate	Actual weight
pencil		
book		
shoe		
person		

Tasks

1 What is the reading on this scale?

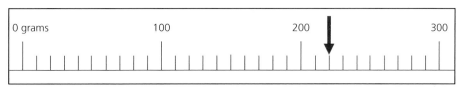

    a 12 g

    b 210 g

    c 220 g

    d 380 g

2 What is the reading on this scale?

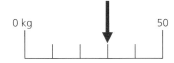

    a 3 kg

    b 6 kg

    c 30 kg

    d 40 kg

3 How many grams (g) are in 1 kilogram (kg)?

4 What is the reading on this scale?

    a 20 g

    b 40 g

    c 200 g

    d 400 g

5 Ben wants to make pancakes. He needs 150 grams of flour.

Ben weighs some flour. The pointer on this scale shows the weight.

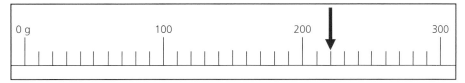

What should Ben do to get the right amount of flour?

Measuring and estimating capacity

Capacity is a measure of the amount a container can hold.

We need to measure capacity accurately in many situations:

- measuring liquids for cooking
- knowing how much liquid there is in a container, for example a bottle of milk
- knowing how much space there is in a freezer or the trunk of a car.

We measure capacity in litres.

A litre is the amount of fruit juice in a standard carton.

A litre can be written as ℓ, so $2\,\ell$ is short for 2 litres.

For smaller amounts, we use millilitres.

A millilitre is a very small amount, roughly the same as a few drops of water.

A millilitre is often written as ml, so 100 ml is short for 100 millilitres.

There are 1000 millilitres in 1 litre.

We use millilitres for smaller amounts and litres for larger amounts.

Here are the capacities of some common containers.

Item	Capacity
Teaspoon	5 ml
Glass of water	200 ml
Can of drink	330 ml
Bottle of water	1 ℓ
Bucket	10 ℓ
Car petrol tank	50 ℓ
Bath	80 ℓ

Using scales

Using measuring instruments means reading a scale.

Example 1

This diagram shows the sort of scale found on measuring jugs. How many millilitres of liquid are in the jug?

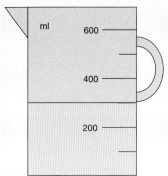

Solution

The level of the liquid is between 200 ml and 400 ml.

There are 2 divisions between the labelled measures of 200 ml and 400 ml.

Those 2 divisions cover 200 ml.

So, each division is 100 ml. ⟵

> 100 ml more than 200 ml is **300** ml – that is the amount of liquid in the jug.

Example 2

This diagram shows a measuring jug with a different scale. How many millilitres of liquid are in the jug?

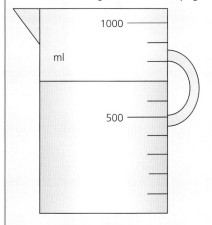

Solution

The level of the liquid is between 500 ml and 1000 ml.

There are 5 divisions between the labelled measures of 500 ml and 1000 ml.

Those 5 divisions cover 500 ml.

So, each division is 100 ml.

The level of the liquid is 2 divisions above 500 ml.

500 + 200 = 700

The amount of liquid is 700 ml.

Example 3

This is a scale on a car fuel gauge. How much fuel is in the car?

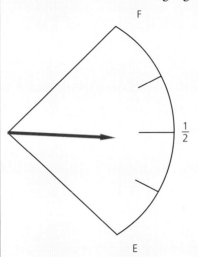

On this scale, F means Full, E means Empty and $\frac{1}{2}$ is half full.

Solution

A fuel tank usually holds 50 litres.

The pointer is showing just under half full.

Half of 50 is 25.

There are no smaller divisions to show the exact amount, so we have to estimate.

There are about 24 litres of fuel.

In different measuring containers, the divisions might not be 100 ml.

They could be 10 ml, 1 litre, 10 litres etc.

Activity

Work out what each division is on these scales.

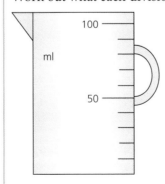

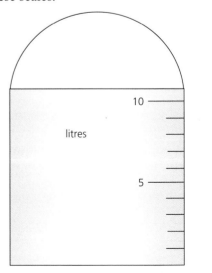

You could use a measuring jug similar to the first one above but it is probably more precise to use a 5 ml teaspoon. You would need to measure two teaspoons of medicine.

How would you measure 10 ml of medicine?

Example

A chef needs 500 ml of milk to make a sauce.

He pours some milk into a measuring jug.

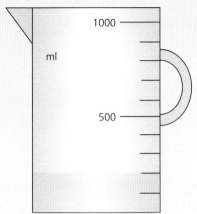

How much more milk should he pour into the jug?

Solution

The level is between 0 and 500 ml.

There are 5 divisions between 0 and 500 ml, so each division is 100 ml.

The level is 2 divisions above 0.

$2 \times 100 = 200$

There are 200 ml of milk in the jug.

He needs 500 ml.

$500 - 200 = 300$

So he should pour 300 ml more milk into the jug.

Tasks

1 Work out what each division is on this scale.

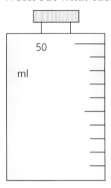

2 How much liquid is there in this bottle?

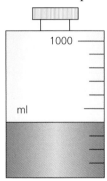

3 How much liquid is there in this container?

4 How much liquid is there in this jug?

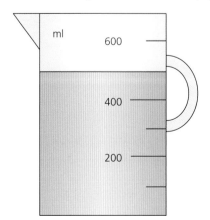

5 Lola is painting a room.

She starts with a tin of paint with 10 litres in it.

This diagram shows the tin of paint when she has finished the room.

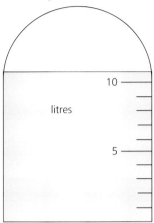

How much paint did Lola use to paint the room?

Measuring temperature

Temperature is a measure of how hot things are.

We need to measure temperature accurately in many situations:

- finding body temperature to see if a person is well
- finding the right temperature for storing and cooking food safely
- finding air temperature so that we wear suitable clothes.

Temperature is measured in **degrees Celsius**, often written as °C, or in **degrees Fahrenheit**, often written as °F.

90 °F means 90 degrees Fahrenheit.

32 °C means 32 degrees Celsius.

Here are some common temperatures.

Description	Fahrenheit	Celsius
Water freezes at	32 °F	0 °C
A typical room temperature is	72 °F	22 °C
Normal human body temperature is	between 98 °F and 99 °F	37 °C
Water boils at	212 °F	100 °C

Measuring temperature

We use a thermometer to measure temperature.

There are many types of thermometer.

A digital thermometer shows the temperature in figures. For example, this thermometer shows human body temperature.

An analogue thermometer shows the temperature on a scale.

Usually there is no pointer on the thermometer. There is a column of liquid and the top of the column shows the temperature.

To read temperature you look at the top of the column and the scale on the thermometer. You need to check if the scale is in °F (Fahrenheit) or °C (Celsius).

Example 1

What is the temperature on this thermometer?

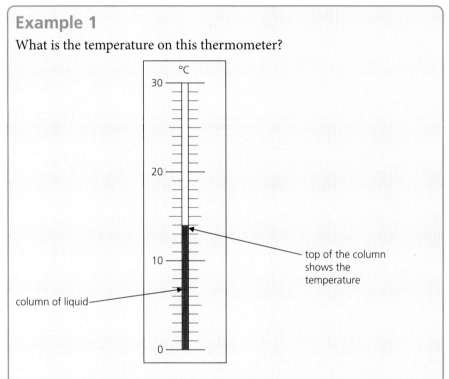

Solution

The top of the column on this thermometer is between 10 degrees and 20 degrees.

There are 10 divisions between 10 degrees and 20 degrees.

Those 10 divisions cover 10 degrees.

So, each division is 1 degree.

The top of the column is 4 divisions above 10 degrees.

The temperature is 14 degrees Celsius or 14 °C.

Example 2

This diagram shows a different thermometer. What temperature does it show?

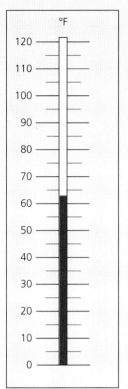

Solution

The top of the column is between 60 degrees and 70 degrees.

There is 1 division between 60 degrees and 70 degrees.

So, each division is 5 degrees.

The temperature is 63 degrees Fahrenheit or 63 °F.

Activity

1 Use a thermometer to find the temperature in your classroom.

2 Use a thermometer to find the temperature outside.

Tip for assessment

Always write the units with your answer.

You may lose marks for missing units.

OVEN TEMP (°F)

Example 3

This is a dial on an oven.

You turn the dial to set the oven temperature.

The pointer shows the temperature. What temperature is the oven set at?

Solution

The pointer is between 350 degrees and 400 degrees.

There are 2 divisions between 350 degrees and 400 degrees.

Those 2 divisions cover 50 degrees.

So, each division is 25 degrees.

The pointer is 1 division (25 degrees) above 350 degrees.

The oven temperature is set at 375 degrees Fahrenheit or 375 °F.

Tasks

1 What is the temperature on this thermometer?

2 What is the temperature on this thermometer?

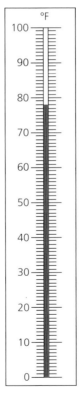

3 What is the temperature on this thermometer?

4 A man measures his own temperature. It is 100 °F. Is he.

    a too cold

    b just right

    c too hot

5 Margaret wants to cook a meal in the oven. The oven temperature in the recipe is 350 °F.

Margaret sets the oven temperature on this dial.

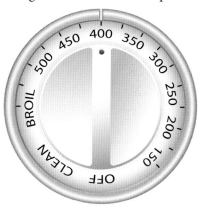

What is the temperature setting on the dial?

Compare it with the temperature in the recipe.

Make one comment.

1 What is the length shown by the pointer?

a 3 cm 20 mm

b 32 cm 7 mm

c 32 cm 8 mm

d 33 cm 2 mm

2 What is the length shown by the pointer?

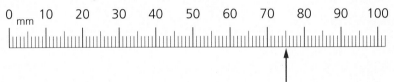

a 75 mm

b 85 mm

c 75 cm

d 85 cm

3 You are facing North and turn a half turn anticlockwise. Where are you facing now?

a North

b East

c South

d West

4 What is 16:30 expressed as a 12-hour clock time?

a 4.30 a.m.

b 6.30 a.m.

c 4.30 p.m.

d 6.30 p.m

5 Your aunt says she will come and see you on the second Thursday in December. What date is this?

December 2019

Monday	Tuesday	Wednesday	Thursday	Friday	Saturday	Sunday
						1
2	3	4	5	6	7	8
9	10	11	12	13	14	15
16	17	18	19	20	21	22
23	24	25	26	27	28	29
30	31					

   a 2 December

   b 5 December

   c 9 December

   d 12 December

6 What time does this clock show?

   a 20 minutes to 4

   b 20 minutes past 7

   c 20 minutes to 5

   d 20 minutes past 8

7 Which is the most likely weight for an aspirin tablet?

   a 100 kg

   b 1 kg

   c 100 g

   d 1 g

8 What is the reading on this scale?

   a 3 kg

   b 6 kg

   c 30 kg

   d 40 kg

9 How much liquid is there in this bottle?

a 7 ml

b 14 ml

c 70 ml

d 97 ml

10 What is the reading on this scale?

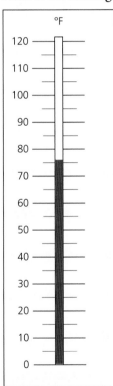

a 75 °F

b 76 °F

c 84 °F

d 86 °F

Unit 103
Pictograms, tables, charts and graphs

Introduction

This unit is about being able to read and understand the information that we meet every day. We see lists, charts, tables and diagrams in lots of places, from the weather chart at the end of a news bulletin to apps on a mobile phone. They all present a summary of information in a visual way, which should be easy to understand!

Learning objectives

In this unit you will find four different ways of presenting information:

- pictograms
- tables (including lists)
- bar charts
- simple graphs

This will help you to prepare for questions about:

- information presented as pictograms, tables, bar charts and simple graphs.

It will also help you to construct your own presentations to show information in different ways.

This unit also covers common banking documentation.

Pictograms

You will be familiar with lots of diagrams used to present information, for example diagrams in instruction manuals, scale drawings and pictograms. At Stage 1 you need to be able to read and draw pictograms, so let us look at these in more detail.

The most important thing to look at is the key because this tells you how many each picture in the pictogram is worth. In the example below, each 😊 represents one child.

Key: 😊 means 1 child

Fruit	Favourite fruit
golden apple	😊 😊 😊 😊 😊
banana	😊 😊 😊 😊 😊 😊 😊
passionfruit	😊 😊 😊
mango	😊 😊
papaya	😊 😊 😊 😊

To find, for example, the number of children who said a banana was their favourite fruit, count the number of faces next to the banana (7) and multiply by the number of children each picture represents (1).

7 × 1 = 7, so 7 children chose bananas.

The picture does not always represent 1; in the example below, each whole apple represents ten apples.

Varieties of apples in a shop	
Red delicious	🍎 🍎 🍎
Golden delicious	🍎 🍎 ◗
Red Rome	🍎 🍎 🍎 🍎
McIntosh	🍎 🍎
Jonathan	🍎 🍎 🍎 ◗

🍎 = 10 apples   ◗ = 5 apples

> To find, for example, the number of Red Rome apples in the food store, count the number of apples next to Red Rome (4) and multiply by the number of apples each picture represents (10). 4 × 10 = 40, so there were 40 Red Rome apples in the food store.

Why is there a picture of half an apple?

> A whole apple represents ten and half of 10 is 5. So the picture of half an apple is worth five apples.

An examination question may simply ask you how many of a certain type of drink were bought or, alternatively, it may ask you to compare the totals of different items bought.

Let us look at an example:

Example

The pictogram shows the number of drinks bought from a sports centre's vending machine on a Monday evening.

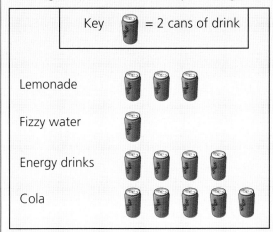

How many more cans of cola were bought from the vending machine than cans of lemonade?

Solution

Step 1: look at the key to find out the number of cans each picture represents (2)

Step 2: use the method given above to find the total number of each type of cans bought (5 × 2 = 10 cans of cola and 3 × 2 = 6 cans of lemonade)

Step 3: take the smaller number away from the larger number to find the difference (10 − 6 = 4 cans)

There were 4 more cans of cola bought than cans of lemonade.

Activity

1 Carry out a survey of students in your class. You may choose to find out how they travel to school, what their favourite food is (you may like to give them four or five choices), what they like to drink or a survey of your own.

2 Once you have the information, decide on a suitable picture (perhaps a plate if you are looking at favourite food) and look at the numbers you have. Can you draw one picture for each choice or have 20 students chosen chicken? You could draw 20 plates but you would need a large piece of paper. It would be more convenient to use a scale.

Choose a suitable scale for the numbers you have.

If you choose a scale of 1 plate to represent 2 students, how many plates would you need? 20 divided by 2 equals 10. So you would need 10 plates.
If you choose a scale of 1 plate to represent 3 students, how many plates would you need? 20 divided by 3 equals 6 remainder 2. You could use this but pictograms only usually have one-half or one-quarter and you would need 6 whole plates and $\frac{2}{3}$ of a plate.
If you choose a scale of 1 plate to represent 4 students, how many plates would you need? 20 divided by 4 equals 5. So you would need 5 plates.
If you choose a scale of 1 plate to represent 5 students, how many plates would you need? 20 divided by 5 equals 4. So you would need 4 plates.
If you choose a scale of 1 plate to represent 10 students, how many plates would you need? 20 divided by 10 equals 2. So you would need 2 plates.

3 Draw your pictogram and remember to show the scale (how many each picture represents).

4 Next, write some questions for your partner to answer.

Tables

A list is a type of table and usually shows only two pieces of information. For example, a shopping list with the number of each item you need to buy.

The first thing to do with any list or table is to read it and work out how it presents the information.

Let us look at an example where you need to use the information in a price list.

Example

The price list shows the cost of different snacks. Which three different types of snack could you buy with $5?

Vanilla cookies: $4.39

Wholewheat crackers: $1.98

Butter cookies: $0.55

Nacho cheese crackers: $0.60

Sugar-free oatmeal cookies: $4.40

Solution

Step 1: choose three items (wholewheat crackers ($1.98), butter cookies ($0.55) and nacho cheese crackers ($0.60))

Step 2: add the cost of the items chosen ($1.98 + $0.55 + $0.60 = $3.13)

Step 3: check that the total is less than $5

Yes, $5 is more than $3.13 ($5 − $3.13 = $1.87)

Although it is tempting to pick the snack that you like best, when you have a limit it is wiser to pick cheaper items first so that you do not go over the allowed amount.

Tip for assessment

Make sure you choose the number of items you are asked for (**3**) and add up the costs to make sure they do not cost more than you are allowed (**$5**).

If, for example, we had chosen the first three items: vanilla cookies (**$4.39**), wholewheat crackers (**$1.98**) and butter cookies (**$0.55**), we would have spent **$4.39 + $1.98 + $0.55 = $6.92** which is too much.

In this example there was only one option that would work but sometimes there may be other options. You can just try different combinations until you find one that works, if you need to.

On shopping sites you will often see more information than we had in the list above. In this table you can see the weight of the packets.

Snack item	Weight (g)	Price ($)
Vanilla cookies	480	4.39
Wholewheat crackers	284	1.98
Butter cookies	55	0.55
Nacho cheese crackers	50	0.60
Sugar-free oatmeal cookies	227	4.40

Tables are often easier to read than lists as the information is structured in columns and rows. You may be asked to construct a table, so you need to look at the information and think of headings for your table.

The top row shows you what is in the table. You know there will be a list of snack items, their weight will be shown in grams (g) and the price in dollars ($). There is no need to show g after each weight or $ before each price as this is in the heading row.

The vanilla cookies don't look so expensive now we can see their weight. We can see there are 480 g for $4.39 so by weight they are the cheapest snack (less than 1 cent per gram). Butter cookies look like the cheapest as the pack is the cheapest, but they are 55 g for $0.55 (1 g for 1 cent).

Which snack is the most expensive?

Sugar-free oatmeal cookies are the most expensive. The pack is only slightly more expensive than vanilla cookies but you get less than half the weight.

Twelve people were asked their favourite colour.

Here is the information presented as a list:

Red

Blue

Red

Green

Red

Blue

Green

Blue

Blue

Blue

Red

Blue

What could the headings for your table be?

Colour and Number would show the information you have.

Tally charts

You need to be able to count methodically and record the number you counted. A good way to do this is in the tally format. A **tally** is where you write down a stroke for each item you count but for the fifth item counted, you 'cross' the four lines you have drawn to make a gate shape, like this:

Each 'gate' is worth five, so you can count in fives.

is equal to 5 + 5 + 2 = 12.

If you are given a list of items and asked to create a tally, it is a good idea to cross each one off as you count.

So if you are given a list like this one:

Red	Green	Green	Blue
Blue	Red	Blue	Red
Red	Blue	Blue	Blue

count each colour in turn, crossing them off as you count, like this:

~~Red~~	Green	Green	Blue
Blue	~~Red~~	Blue	~~Red~~
~~Red~~	Blue	Blue	Blue

and complete the tally chart:

Colour	Tally	Number of people
Red	IIII	4
Green	II	2
Blue	HHH I	6
Total		12

Activity

Complete another survey, or use the information you collected in the pictogram activity, and make a table to show your results.

Bar charts

Newspapers, brochures and trade magazines often use charts to present information. At Stage 1 you need to read, understand and draw **bar charts**.

Bar charts present information as a series of blocks or bars. The height of each bar represents the amount for that particular category.

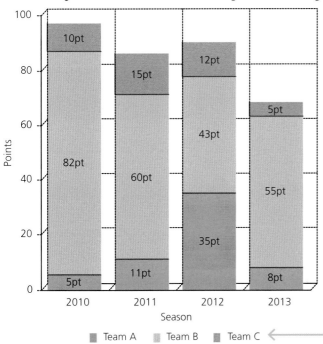

▶ Scores by team per season

The lines with numbers or names going across and up the page are called the **axes**. The line going across the page is the **horizontal axis** (imagine the horizon lying across the sea) and the one going up is called the **vertical axis**.

If different colours are significant, there will be a key to show you what each colour means. (The key is sometimes called a legend.)

Bar charts can have the category names going up and the number scale going across. This does not change what you have to do – the chart is just turned on its side and the length of the bar represents the amount for each category.

Learner tip

If you are asked to draw a bar chart in the exam, always remember to add a title, a key and a label for both the axes.

Tip for assessment

Work out the scale on the vertical axis first so that you know what each box or line on the bar chart represents. In this example, the vertical scale goes up in $10 increments, so halfway between any two labelled lines is an increase of $5.

The first thing you need to look at on a bar chart is the vertical scale (the scale that goes up the side of the chart). This will help you read any information that you need from the chart accurately. The scale on the vertical axis of a bar chart always starts at 0.

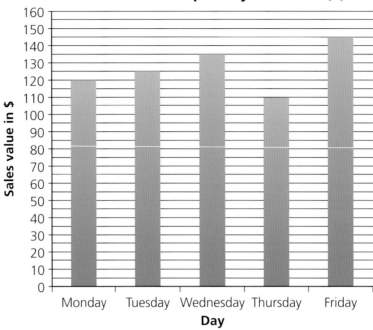

Ice-cream sales per day in dollars ($)

There are several types of exam question you could be asked about this bar chart. Let us look at an example of each:

1 What was the total value of ice-creams sold at the trade fair on Monday?
2 What is the difference between the value of ice-creams sold on the Tuesday and the Friday?
3 Which day had the lowest amount of ice-cream sales?

To find the answers you need to read across from the day you are being asked about to see where it is on the scale.

1 On **Monday** the sales of ice-creams were worth **$120**.
2 To find the difference:
   • Step 1: find the value of sales on the given days (**Tuesday $125** and **Friday $145**)
   • Step 2: find the difference between the two (**$145 – $125 = $20**)
   So, the answer is **$20**.
3 Look for the shortest bar: the day with the lowest value of ice-cream sales is **Thursday** with **$110**.

Activity

Use the information you collected in the pictogram activity, or the table showing favourite colours, to draw a bar chart. Remember to add a title and label the axes.

Line graphs

Line graphs are usually used to display continuous data and are frequently used by businesses and other organisations to show data about production, sales turnover and profit/loss.

A line graph has horizontal and vertical **axes** like a bar chart. A line graph should have a meaningful title and both axes should be labelled clearly. The vertical scale does not have to start at 0.

This example shows the value of sales each month. June has the highest value of sales and April has the lowest value.

You could be asked to draw or extract information from a line graph.

Drawing a graph

Let us look an example of drawing a line graph.

Draw a line graph for the following data:

▼ Growth of a plant

Monday	Tuesday	Wednesday	Thursday	Friday	Saturday	Sunday
20 cm	23 cm	28 cm	32 cm	39 cm	44 cm	50 cm

Step 1: choose your scale for the vertical axis (Consider the highest (50) and lowest numbers (20, so use 0)

Step 2: plot the value for each day at the correct height

Step 3: join each point to the next with a straight line

Step 4: label the axes and add a suitable title

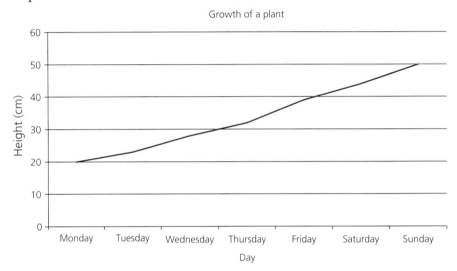

Extracting information from a graph

Now let us look at an example of extracting information from a graph.

Use the conversion graph to convert 20 US Dollars to Jamaican Dollars.

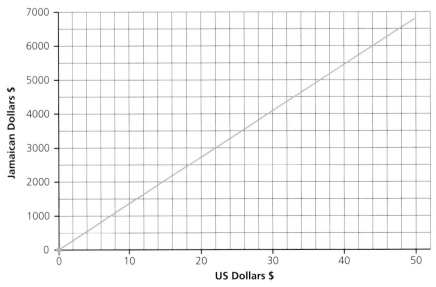

▲ Conversion graph for Jamaican Dollars and US Dollars (October 2018)

Step 1: find the value that you need ($25) on the horizontal axis and follow it up until you reach the line of the graph

Step 2: follow it across to the vertical axis and read the value at this point ($3400).

So, an item which costs US $25 will cost 3400 Jamaican Dollars.

Activity

Draw a line graph to show this information:

▶ Sales July to December 2018

Month	Sales ($)
July	350
August	400
September	380
October	390
November	410
December	420

Remember to choose a suitable scale and label your axes.

Read common banking information

For the examination, you may be asked to read simple banking documentation and recognise some banking terms. Banking documentation is similar to a list or table.

On the next page is an example of a **deposit slip** (also known as a lodgement form). You would use this to pay cash and cheques into a bank or building society.

First you sort the money you are depositing so you have all the same denomination (same value) notes together (for example all the $100 notes in one pile and the $50 dollar notes in another pile).

Then you count the number in each pile and list them on the form.

LODGEMENT		
ANY BANK DATE _____/_____/_____ (DAY) (MONTH) (YEAR)		✕ 5,000.00 ————————————
		✕ 1,000.00 ————————————
BRANCH _____		✕ 500.00 ————————————
ACCOUNT NO. _____		✕ 100.00 ————————————
NAME OF ACCOUNT HOLDER _____		✕ 50.00 ————————————
		✕ 20.00 ————————————
PAID IN BY _____		✕ 10.00 ————————————
TELEPHONE NO. _____		✕ 5.00 ————————————
SOURCE OF FUNDS _____		✕ 1.00 ————————————
		COINS ————————————
(See reverse for details if necessary) _____		CHEQUES TOTAL _____ (list on back if necessary)
		LESS CASH RECEIVED _____ (from proceeds of cheques)
CURRENCY		_____
JA$ ◯ US$ ◯ CDN$ ◯ GBP£ ◯		TOTAL
		SIGNATURE _____

This is an extract from the form above. In this example we have seven $100 notes and five $50.

> 7 × 100 = 700, so we enter this on the form.
>
> 5 × 50 = 250, so we enter this on the form.
>
> If this is all we are paying in, we can total this: 700 + 250 = 950.
>
> The total amount to pay in is $950.00 We put the zeros in the cents column on banking documentation to show there are no cents and we haven't just forgotten them.

→

	×	5 000.00	
	×	1 000.00	
	×	500.00	
7 ×		100.00	700.00
5 ×		50.00	250.00
	×	20.00	
	×	10.00	
	×	5.00	
	×	1.00	
	Coins		
	Cheques		
	Total		950.00

Bank statements

Your **bank statement** shows the amount in your account at the start of a period and what money has gone into and gone out of your account during the period. This gives a final balance of the amount in your account on the date of the statement. The final balance is known as a Closing balance.

Any Bank

Account Transaction History
Wednesday 26 September 2018

Account: 00123456

Opening balance: $343.50   Closing balance:

Posting date	Reference number	Description	Cheque number	Additional details	Debit	Credit	Running balance
02-09-18	TTT34567	Cash withdrawal			$200.00		$143.50
16-09-18	WW3256	Debit re purchase			$90.00		$53.50
21-09-18	MA2345	Cash paid in					

In the example above, the Opening balance is shown above the columns. This is the amount you have in your account at the start of the statement period.

The opening balance is $343.50.

On 2 September cash was taken out of the account and shown in the Debit column. This is shown as Cash withdrawal under Description.

Any money taken out of the account is shown in the Debit column.

The Running balance is $343.50 − 200.00 = 143.50$

On 16 September money was taken out of the account for a purchase. This is shown as a Debit in the Description column (whereas when you took cash out on 2 September this was shown as cash withdrawal). A debit here is where the other person is able to take the money out of your account because you have previously signed to say they can do this. Direct debits are often used for loan payments and other regular payments.
$143.50 − 90.00 = 53.50$

On 21 September you paid in the dollars you entered on the deposit slip earlier.

What would the entry for this look like on the bank statement?

How much money will be in the account after this money is paid in?

21-09-18	MA2345	Cash paid in				$950.00	$1003.50

The Description column shows Cash paid in and the amount ($950.00) is shown in the Credit column. Any money paid into the account is shown in the **Credit** column.

The Running balance is now $1003.50.

If this was the last amount paid in or out of your account before the date given on the statement, this would also be the Closing balance and be shown at the top of the statement.

Money can go into your account in different ways:

- cash deposit (using a deposit slip similar to the example above)
- incoming wire transfer
- interest
- cheque paid in.

Money going in to your account is shown in the Credit column.

Money can go out of your account in different ways:

- debit payment
- cash withdrawal
- bank charges (such as incoming wire charges)
- cheque payment.

Money going out of your account is shown in the Debit column.

Example

This is a different customer's account.

Any Bank

Account Transaction History
Monday 1 October 2018

Account: 00123458

Opening balance: $908.00   Closing balance:

Posting date	Reference number	Description	Cheque number	Additional details	Debit	Credit	Running balance
02-09-18	RTT34567	Cash paid in					
16-09-18	SW3256	Bank charge					
21-09-18	TA2345	Cash withdrawal					
28-09-18	VWW433	Cheque paid in	345678				

Record the following payments in and out of the account.

a Cash paid in $ 30.00
b Bank charge $ 0.10
c Cash withdrawal $100.00
d Cheque paid in $342.60
e What is the closing balance?

→

Solution

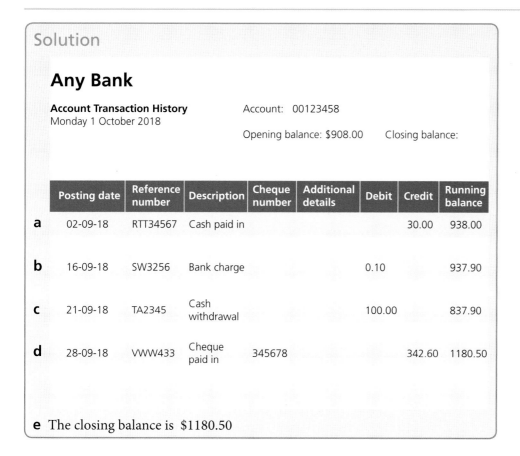

Any Bank

Account Transaction History   Account: 00123458
Monday 1 October 2018

Opening balance: $908.00   Closing balance:

	Posting date	Reference number	Description	Cheque number	Additional details	Debit	Credit	Running balance
a	02-09-18	RTT34567	Cash paid in				30.00	938.00
b	16-09-18	SW3256	Bank charge			0.10		937.90
c	21-09-18	TA2345	Cash withdrawal			100.00		837.90
d	28-09-18	VWW433	Cheque paid in	345678			342.60	1180.50

e The closing balance is $1180.50

Tasks

1 The pictogram shows the drinks chosen by staff last Tuesday.

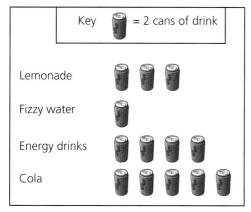

Key   = 2 cans of drink

Lemonade
Fizzy water
Energy drinks
Cola

  a How many cans of lemonade were chosen?
  b What was the most popular drink?
  c How many more cans of cola were chosen than energy drinks?

2 Look at this pictogram.

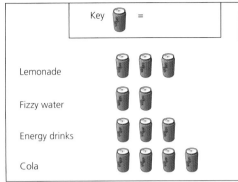

If six cans of fizzy water were bought from the machine, how many cans does each picture represent?

3 You are drawing a pictogram and want to use this image to represent 5 workers.

How many of them would you need to represent 40 workers?

4 Look at the shop opening times below.

Shop Opening Times

Monday	Closed
Tuesday	9 a.m.–5 p.m.
Wednesday	9 a.m.–1 p.m.
Thursday	9 a.m.–8 p.m.
Friday	9 a.m.–6 p.m.
Saturday	9 a.m.–6 p.m.
Sunday	10 a.m.–4 p.m.

a At what time does the shop open on a Friday?

b At what time does the shop close on a Thursday?

c On how many days does the shop open at 9am?

d For how many hours is the shop open on a Sunday?

5 Copy and complete the tally table for these coloured circles.

Colour	Tally	Total in figures
Red ●		
Yellow ○		
Blue ○		
Purple ●		
Green ○		

6 Draw a bar chart to represent the information in Question 5.

7 The bar chart shows how long it took for Helen to get to work each day.

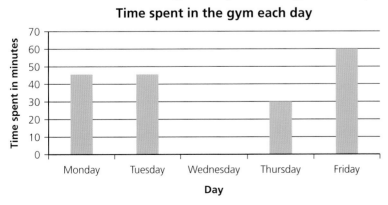

Time taken to get to work each day in minutes

a On which day did it take the longest for Helen to get to work?

b How long did it take Helen to get to work on Wednesday?

8 The bar chart shows how long Aisha spent in the gym each day.

Time spent in the gym each day

a On which day did Aisha not go to the gym?

b On which day did Aisha spend longest in the gym?

9 Use this graph to find out how many centimetres there are in 20 inches.

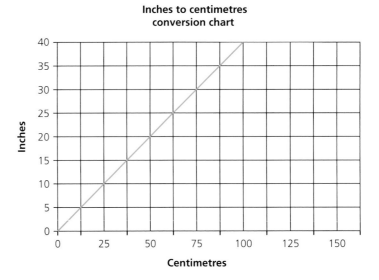

Inches to centimetres conversion chart

10 Any Bank

Account Transaction History
Monday 2 January 2019

Account: 00199999
Opening balance: $250.00   Closing balance:

	Posting date	Reference number	Description	Cheque number	Additional details	Debit	Credit	Running balance
a	02-12-18	STT34567	Cheque	785678				
b	16-12-18	RW3256	Cash paid in					
c	21-12-18	VA2345	Interest					
d	28-12-18	FWW433	Cash withdrawn					

Use a copy of the bank statement above to record the following payments in and out of the account.

a Cheque (you have written) $76.00

b Cash paid in $30.00

c Interest received $0.20

d Cash withdrawal $100.00

e What is the closing balance?

1 A student wants to use this symbol in his pictogram recording the amount people spend on phone calls.

$ represents $5

How many of the $ symbols should he use to represent $40?

a 4

b 5

c 8

d 10

2 The pictogram shows the number of drinks bought from a vending machine.

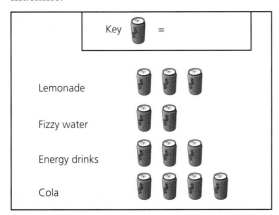

If 15 cans of energy drink were bought from the vending machine, what number of cans should each picture represent in the key?

a 6

b 5

c 3

d 2

3 The weather chart shows the forecast for the week.

Monday	Tuesday	Wednesday	Thursday	Friday
☀	☁	🌧	☀	⛈

What is the weather forecast for Wednesday?

a sun

b thunderstorms

c rain

d cloud

4 The menu below shows the price of various fish dishes.

Entree	Price
Fried fish	$1200
Steamed fish	$1200
Brown stewed fish	$1300
Curried fish	$1250
Curried shrimp	$1500
Shrimp in garlic	$1550
Curried lobster	$1650

How much does curried fish cost?

a $1200

b $1250

c $1500

d $1650

5 The bar chart shows the time, in minutes, that Benito spends travelling to work each day.

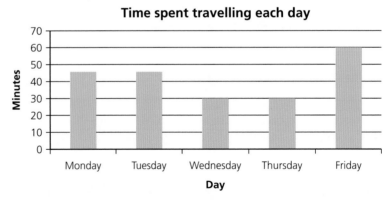

What is the difference between the time spent travelling to work on Tuesday and Wednesday?

a 10 minutes

b 20 minutes

c 15 minutes

d 0 minutes

6 What number do the tally marks below represent?

a 19

b 18

c 14

d 20

7 The graph shows how far a person will walk in different amounts of time.

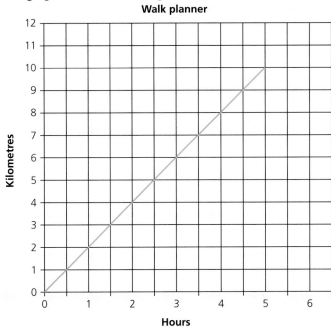

How far can the person walk in 4 hours?

a 2 km

b 4 km

c 6 km

d 8 km

8 A car salesperson records the sales of cars in one month.
The chart is not complete. What is missing?

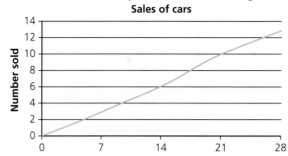

a a title

b a scale on vertical axis

c a label on horizontal axis

d a label on the vertical axis

9 A bank customer takes $50 cash out of her account. On her bank statement this will be a:

a credit

b direct debit

c standing order

d withdrawal

10 This is an extract from a bank paying-in form. How much in total is the customer paying in to her account?

×	5000.00	
×	1000.00	
×	500.00	
3 ×	100.00	
1 ×	50.00	
×	20.00	
2 ×	10.00	
×	5.00	
×	1.00	
	Coins	
	Cheques	
	Total	

a $6

b $300

c $370

d $600

Introduction

Shape is another mathematical concept which is rooted in everyday life. The names of shapes and their features are often quite familiar. Circles and squares are commonly recognised **two-dimensional (2D)** shapes, while cubes and cuboids are examples of **three-dimensional (3D)** shapes.

Does shape actually matter? Try playing football with a cube-shaped ball before making that decision! One phrase you may have heard is 'trying to fit square pegs in round holes'. This suggests that someone is attempting to do the impossible. The use of shape in the phrase helps us to visualise that the task cannot possibly work.

To pass a maths exam, you need to know the names of common two-dimensional (2D) and three-dimensional shapes (3D) and what they look like. For example, you need to know how many corners and sides they have, whether they roll and if they have curved sides.

In days of old, it was widely believed that the world was flat.

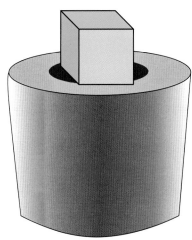

▲ Trying to fit a square peg into a round hole. Impossible!

▲ Is this what would happen if the world was flat?

Assuming that the scientists of the day knew that our planet was round, they must have pictured it as a circle. That would be a 2D shape. Now, of course, we know that the earth is a sphere (globe shaped) and is therefore 3D. One way of telling the difference between a 3D shape (a solid shape you can hold) and a 2D shape (flat) is that a 3D shape has three measurements – length, width and depth – while the 2D shape has only two, length and width.

Learning objectives

In this unit you will find information on:

- measuring the perimeter of shapes such as rectangles and circles
- calculating the perimeter and area of rectangular shapes
- line symmetry
- simple volume
- tessellation (e.g. whether a shape would be suitable for tiling a floor) and working with nets of cubes and cuboids.

Measuring and drawing shapes

The diagrams below show some common shapes and their mathematical names.

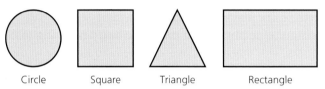

Circle   Square   Triangle   Rectangle

▲ Some 2D shapes

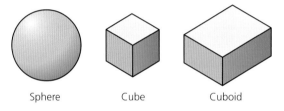

Sphere   Cube   Cuboid

▲ Some 3D shapes

How would you recognise or draw these shapes?

Not only will you need to memorise the names of these common shapes but you will also need to know some of their properties. This means the number of sides they have and the number of corners.

You will also need to master some shape vocabulary: In a 2D shape, a corner is known as a **right-angle**. In a 3D shape like a cuboid the sides, including the top and bottom, are known as **faces**.

A **rectangle** looks easy to draw but you need to know that it has 4 sides (in the diagram above you can see 2 long sides, 2 shorter sides) and 4 right-angles (its corners).

What are the properties of a **square**? A square is a special type of rectangle. It has 4 sides and 4 right-angles but the sides are all the same length.

A ruler and pencil.   ⟶   What do you need to draw a rectangle or square?

You need a ruler to check that the opposite sides are the same length in a rectangle or all the sides are the same length in a square.

How can you draw a circle?

You can't draw along a ruler as this gives straight lines. You could draw around a coin or a plate but to draw circles of different sizes you need a pair of compasses. You can use your ruler to measure the distance you need for your compasses.

How can you measure the distance around a shape?

If the shape has straight sides, like a rectangle or square, you can use a ruler. If not, you can use something flexible, like a piece of string, to go around the shape and then measure the piece of string.

Perimeter

The **perimeter** is the length of measurement around an entire two-dimensional (2D) shape. Think of the perimeter as the surround, or boundary, of these shapes where the total measurements of each width and length are added together. If, for example, a yard is rectangular then by adding the measurement of each side (two lengths and two widths) you would be able to calculate the total length of fencing required to border the perimeter of the yard.

One way to remember what the perimeter of a shape means is to think of

- p for path – a path around the shape and
- meter is another spelling for metre which is a measurement of length.

Here is a rectangle.

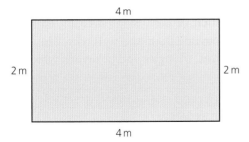

What is the perimeter of this shape?

If the shape was drawn to the correct size we could measure around the shape, but the measurements show this is 4 metres long and 2 metres wide. We can add up the measurements

$4 + 2 + 4 + 2 = 12$

So the perimeter of this shape is 12 metres.

What is the perimeter of this shape?

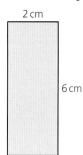

2 cm

6 cm

> There are 4 sides so we need 4 measurements.

For the perimeter we add up the measurements of all the sides.

As this is a rectangle, we know the longer sides are both the same length and the shorter sides are both the same length. So the perimeter is 6 + 2 + 6 + 2 = 16. Look at the measurements on the diagram. They are in cm. So the perimeter is 16 centimetres.

Real world maths

A builder uses measurements when constructing buildings. Can you imagine if a builder guessed the sizes of doors and windows rather than measuring the appropriate space when building?

Activity

1 Find the perimeter of your table top.
2 Find the perimeter of the classroom floor.

Learner tip

Remember you don't need to measure all the way round a rectangle as long as you know the measurements of one of the longer sides and one of the shorter sides.

Don't forget the units. Have you measured in metres, centimetres or millimetres?

Tessellation

We say that shapes **tessellate** when they fit together leaving no gaps. Good examples of practical shape tessellation include floor and wall tiles. The following images show examples of squares, triangles and hexagons which all tessellate.

▲ These shapes all tessellate.

Circles, as the example below demonstrates, do not tessellate as they cannot fit together without leaving gaps.

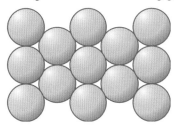

▲ Circles do not tessellate.

Activity

Will a rectangle tessellate?

Check your answer by trying rectangles of different sizes.

Area

The **area** of a 2D shape is the space taken within a flat surface.

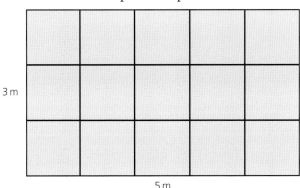

3 m

5 m

This shape could be a room and we need to know the area to find out how much floor covering to buy. The room in the diagram has been divided into square metres. Each square inside the shape is 1 metre × 1 metre. We could count the number of squares = **15**.

Or we can multiply the length and width together. For a room measuring 5 m (length) by 3 m (width), the area would therefore be $5 \times 3 = 15 \text{ m}^2$.

The small '2' above the m represents squared, showing that there is an overall area consisting of 15 squares, with each square measuring 1 metre on all four sides.

Find the area of this square.

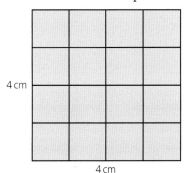

4 cm

4 cm

To find the area we can multiply length and width together.

The area is $4 \times 4 = 16$. This time the measurement is centimetres, so the area is 16 cm^2.

Activity

Draw a rectangle 4 cm by 2 cm.

Now calculate the area by multiplying.

Check your answer by drawing in the squares and counting them.

Making 3D shapes from nets

3D shapes can be created by folding 2D shapes. These 2D shapes are called **nets** and exam questions sometimes ask you which net will make a given 3D shape.

Cuboid

Here is the net of a cuboid.

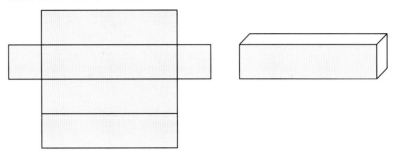

Activity

Trace this shape, or draw one of your own, and then cut it out and fold along the lines to make a cuboid.

The cuboid you made is like a box with a lid. It shows all 6 faces of a cuboid.

Here are a few other examples you may see.

Square-based pyramid

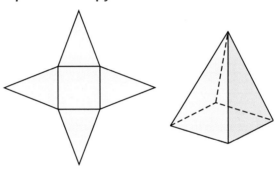

Triangular prism

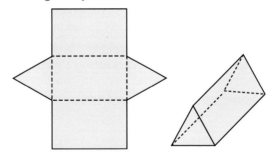

> **Tip for assessment**
>
> Count the number of faces to see if there are sufficient for the given shape. For example, for a cube or cuboid you will need 6 unless the question says it is a box without a lid.
>
> Next check if the faces are the correct size and shape. For example, if you are making a cube all the faces need to be the same size.
>
> Finally, check if the net would fold up to make the given shape.

Volume

We have calculated the length for the perimeter of a shape (e.g. around a room) and the area of a 2D flat surface (e.g. a floor measured in squared units). Now we will look at the volume of a solid shape; these shapes will be 3D, such as a cube or cuboid.

Volume is the space taken by a 3D object. Look at this cuboid.

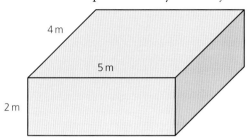

To find the volume we multiply the dimensions of the length, width and height.

Therefore, the volume is $5\,m \times 4\,m \times 2\,m = 40\,m^3$.

Remember, for area we multiplied two measurements (length and width) and the answer was in squared units such as m^2. For volume we have multiplied three measurements (length, width and height) and the small '3' represents cubic metres. This shows that there is an overall volume of 40 square cubes, with each cube measuring 1 m in length, width and height.

Imagining a Rubik's cube can help when trying to understand the concept of volume. Consider the construction of the cube using multiple 1 cm cubes. The bottom tier would require nine cubes (three cubes wide by three cubes long). You would then stack a further two tiers of nine cubes. As each of the three tiers contains nine cubes, you would have used 27 (9 × 3) 1 cm³ cubes. If you measured the volume of your finished cube (length × width × height), the calculation would be
$3\,cm \times 3\,cm \times 3\,cm = 27\,cm^3$

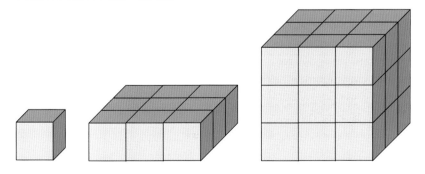

> **Tip for assessment**
> Always remember to indicate whether an answer is squared (²) for area or cubed (³) for volume.

Lines of symmetry

A **line of symmetry** is a line that you can draw through the middle of a shape so that both halves are identical. The shape is divided into two matching halves. Another way to see this line is to put a mirror along the line of symmetry and the half-shape with its reflection will look the same as the original shape. Sometimes there are more lines of symmetry than you expect and sometimes there are none!

Activity

Cut out similar shapes and try it for yourself. Take care that your square has equal length sides and four right-angled corners or else it might not work!

Now try a circle.

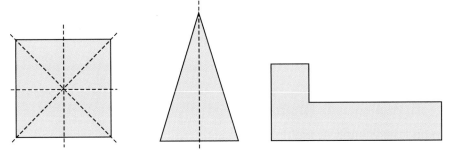

▲ Some shapes have lots of lines of symmetry, others have just one, while some have none at all.

Notice that the square has four lines of symmetry – two which run corner to corner and two which split the shape halfway across and halfway down. The last shape has no lines of symmetry.

A circle has an infinite number of lines of symmetry. You can choose any point on the circle and draw a line through the centre to the other side to get a line of symmetry.

The example below identifies one line of symmetry on a regular hexagon. The line could have been obtained by folding the shape in half.

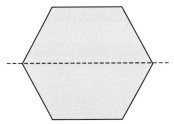

Different lines of symmetry could have been obtained by folding the same shape in different places. A regular hexagon has 6 lines of symmetry. The example below demonstrates three different lines of symmetry in an **equilateral triangle** (equilateral means all the sides are the same length).

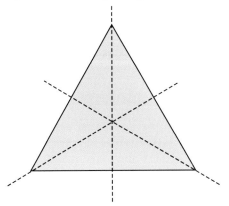

Tasks

1 **a** What is this shape?

  b How many equal angles does this shape have?

  c How many equal sides does this shape have?

2 What would you use to draw a circle?

3 Which of the following is not a 3D shape?

- cube
- rectangle
- cuboid

4 What is the total length of the sides of this rectangle?

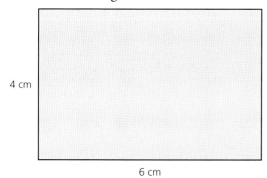

5 What is the perimeter of this shape?

6 What is the area of this shape?

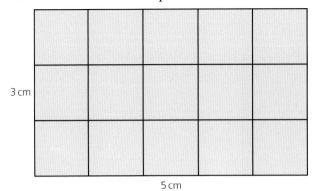

3 cm

5 cm

7 What is the area of a square with sides of 5 cm?

8 Draw a net for a cube.

9 Find the volume of this cuboid.

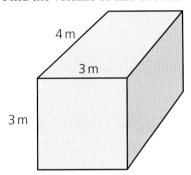

4 m

3 m

3 m

10 Does this shape have any lines of symmetry?

Copy the shape and draw the line or lines of symmetry.

Test your knowledge

1 What is this shape?

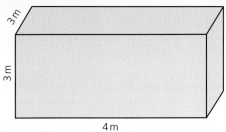

a a cube

b a cuboid

c a cylinder

d a rectangle

2 How many faces does the shape in question 1 have?

    a 3

    b 4

    c 6

    d 8

3 What is the perimeter of this shape?

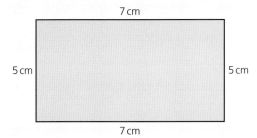

    a 12 cm

    b 24 cm

    c 28 cm

    d 35 cm

4 What is the area of this shape?

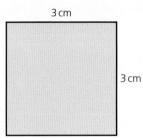

3 cm

3 cm

a 6 cm²

b 8 cm²

c 9 cm²

d 12 cm²

5 Which of the following shapes would be suitable for floor tiles (assuming that you wish your floor tiles to tessellate)?

A   B   C   D

 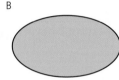 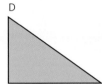

a Shape A only

b Shapes A and C

c Shapes C and D

d Shapes A and D

6 Which of these nets would form to make a cube?

a   **b**   **c**   **d**

7 What is the area of this shape?

8 m

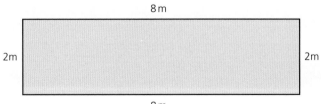

2 m   2 m

8 m

a 10 m²

b 16 m²

c 18 m²

d 20 m²

8 Find the volume of this box.

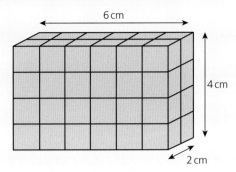

a 12 cm³

b 14 cm³

c 26 cm³

d 48 cm³

9 Which of these lines is a line of symmetry?

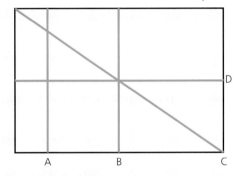

a lines A and B

b lines B and C

c lines C and D

d lines D and A

10 The perimeter of this shape is

a 10 cm

b 10 cm²

c 10 cm³

d 10 cm⁴

Introduction

This unit will help you to develop confidence in adding, subtracting, multiplying and dividing whole numbers.

Improving your skills in adding and subtracting will help you in lots of ways in day-to-day life, for example working out how much your shopping bill should be and how much change you will get if you use a $20 note. You will find it much easier to learn the rest of the maths at this level, and at higher levels too, if you become more confident with these skills.

Even though you may never be without the calculator on your mobile phone, it is still very important to be able to add and subtract quickly in your head and on paper. You need to know if your calculator gives you the wrong answer, for example if you made a mistake when typing in the calculation, and, with practice, it is often quicker to add or subtract in your head. Also, you will not be able to use a calculator to help you in the City & Guilds' examination.

This unit will help you to build on what you know about adding and subtracting numbers to 10 and 20 by showing you simple ways of adding and subtracting larger 2- and 3-digit numbers accurately and quickly, in your head and in writing.

When learning about multiplication, it can sometimes help to think of it as a quick way to do repeated addition. We will see how, as we work through this unit. When you can recall the multiplication tables quickly, you will have the confidence to move onto multiplying and dividing larger numbers.

At this level, you will need to know what multiplication is, how to multiply single-digit numbers (such as 3×7) and understand when this is the right mathematics to use in practical situations. You should be aware of the fact that the words 'times', 'multiply by' and 'lots of' mean the same thing.

It is important to understand that 3×5 gives the same answer as 5×3, but that in a practical situation they may mean different things. For example, 3 lots of 5 kg bags is different from 5 lots of 3 kg bags, but the total number of kilos is the same in both instances, 15 kg.

When learning about division, it may help to think of it as sharing the amount you have. It is similar to repeated subtraction. At this level, you should know and understand that multiplying and dividing are opposite operations. The emphasis in this section is on building knowledge of some of the multiplication tables and beginning to use them when multiplying 2-digit numbers, such as 34×3. You will then use this knowledge to do the opposite: divide 2-digit numbers by single-digit numbers.

Learning objectives

In this unit you will find more information on:

- addition
- subtraction
- multiplication
- division.

Addition and subtraction

It is really important to become confident with adding and subtracting smaller numbers. Make sure that you know all the pairs of numbers that add to 10: 0 and 10; 1 and 9; 2 and 8; 3 and 7; 4 and 6; 5 and 5. It may seem silly, but it will really help you to work out lots of additions and subtractions quickly. Once you know these facts you can work out lots of additions and subtractions easily:

$0 + 1 = 10$	$10 - 0 = 10$	$10 - 10 = 0$
$1 + 9 = 10$	$10 - 1 = 9$	$10 - 9 = 1$
$2 + 8 = 10$	$10 - 2 = 8$	$10 - 8 = 2$
$3 + 7 = 10$	$10 - 3 = 7$	$10 - 7 = 3$
$4 + 6 = 10$	$10 - 4 = 6$	$10 - 6 = 4$
$5 + 5 = 10$	$10 - 5 = 5$	$10 - 5 = 5$

You can use these facts to help you work out additions and subtractions up to 20.

Activity

Put these cards into pairs so that each pair is equal to 20 when added.

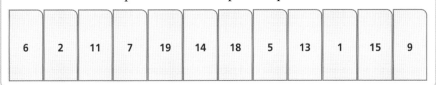

| 6 | 2 | 11 | 7 | 19 | 14 | 18 | 5 | 13 | 1 | 15 | 9 |

Learner tip

Use number pairs to 10 to help you tackle this task. For example, $6 + 4 = 10$, so $16 + 4 = 20$.

Using known facts in addition and subtraction

Knowing number pairs to 10 and 20 can help you add and subtract other numbers too, for example numbers near 10 or 20.

What about 19 − 6?

There is more than one way of working this out. Different methods work for different people and different numbers. Here are three ways:

1 Using known facts to 10
   You know 9 − 6 is 3, so 19 − 6 = 13

2 Using known facts to 20
   19 is one less than 20, so if 6 + 14 is 20, it will be one less than this so 19 − 6 = 13

3 A number line is another useful tool for solving addition and subtraction questions.
   The number line below shows how you can use this method to work out the difference between 6 and 19, that is 19 subtract 6.

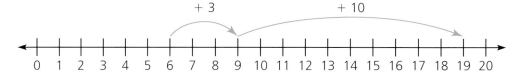

From 6 to 9 you **add** 3 and from 9 to 19 you **add** 10. 10 + 3 = 13, therefore 19 − 6 = 13.

Learner tip

If you have to answer a question mentally, picture a number line to 20 and move up and down it.

Addition and subtraction with 2- and 3-digit numbers

When working with bigger numbers, you may want to use a more formal written method using columns. Don't panic; in the next section, we will work through some examples. The most important thing to remember is to make sure the digits line up: units with the units, tens with the tens and hundreds with the hundreds. If they are not lined up, you may confuse a hundred for ten and get the wrong answer.

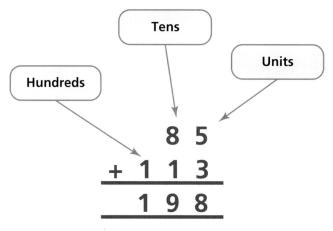

Set the addition in columns. Paper with small squares can be useful to help. Add the digits in each column.
6 + 2 = 8 and 4 tens + 3 tens = 7 tens

Remember to set the addition in columns.

Add the digits in the units column (**8 + 6 = 14**, write **4** in the **units** column and a small **1** under the **tens** column to show you have an extra 10 to add).

Add the digits in the tens column (**3 + 2 = 5, 5 + 1 = 6**). There is only 1 digit in the hundreds column, so write this (**1**).

Here **6 + 6 = 12**, write **2** in the **units** column and a small **1** under the **tens** column to show you have an extra 10 to add). Add the digits in the tens column (**7 + 8 = 15, 15 + 1 = 16**), write **6** in the **tens** column and a small **1** under the **hundreds** column to show you have an extra 100 to add). There are no digits in the hundreds column, so **0 + 1 = 1.**

First set the subtraction in columns.

Then subtract the digits in each column.
(**8 − 2 = 6 and 7 tens − 3 tens = 4 tens**)

First set the subtraction in columns.

Then subtract the digits in the units column.

You start with 4 units but cannot take 8 units from 4 units. You will need to use partitioning and split the number differently. Instead of 100 + 60 + 4, **164 becomes 100 + 50 + 14**. Change the **6 tens to 5 tens** to show that you have moved 1 ten to the units column and change the **4 units to 14 units**. Now you can subtract the units (**14 − 8 = 6**). Write **6** in the **units** column.

Subtract the digits in the tens column (**5 tens − 3 tens = 2 tens**).

Write **2** in the **tens** column.

There is only 1 digit in the hundreds column, so write this (**1**).

Understanding how to partition numbers will help you to add and subtract in your head and using columns. We can think of 146 as 100 + 40 + 6, but if we need to take 8 away from 146, it may help to think of 146 as 100 + 30 + 16.

Addition

Use column addition to work out these.

a 46 + 32

$$\begin{array}{r} 4\ 6 \\ +\ 3\ 2 \\ \hline 7\ 8 \end{array}$$

The answer is **78**.

b 38 + 126

$$\begin{array}{r} 3\ 8 \\ +\ 1\ 2\ 6 \\ \hline 1\ 6\ 4 \\ {\scriptstyle 1} \end{array}$$

The answer is **164**.

c 76 + 86

$$\begin{array}{r} 7\ 6 \\ +\ \ \ 8\ 6 \\ \hline 1\ 6\ 2 \\ {\scriptstyle 1\ \ 1} \end{array}$$

The answer is **162**.

Use column subtraction to work out these.

a 78 − 32

$$\begin{array}{r} 7\ 8 \\ -\ 3\ 2 \\ \hline 4\ 6 \end{array}$$

The answer is **46**.

b 164 − 38

$$\begin{array}{r} 1\ {\scriptstyle 5}6\ {\scriptstyle 1}4 \\ -\ \ \ 3\ 8 \\ \hline 1\ 2\ 6 \end{array}$$

The answer is **126**.

Sometimes you will need to work out whether to add or subtract the numbers to find the answer you need. It doesn't matter which method you use to add or subtract.

Let us look at some examples of real-life problems.

Example 1

A candy cost $16. You pay for it with a $20 note. How much change will you get?

Solution

First you need to work out the sum you need to answer for the question. (You have $20 and spend $16, so the sum is 20 − 16.)

Work out the calculation (you should know your number pairs to 20, so 20 − 16 = 4).

Give your answer in the context of the question ($4).

Example 2

You want to buy a snack which costs $19. You check to see how much money you have and find you have $8. How much more money do you need to buy the snack?

Solution

First, work out the sum you need to answer for the question. (You have $8 and need $19, so the sum is 19 − 8).

Work out the calculation.

You could use a number line for this.

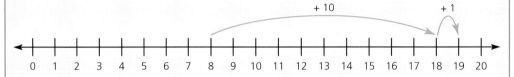

From the number line, 10 + 1 = 11

Give your answer in the context of the question ($11).

Example 3

a You need to cut two pieces of rope: one should be 85 mm and the other 127 mm. How much rope do you need altogether?

b You cut the two pieces of rope from a longer rope which is actually 280 mm in length. How much rope is left over?

Solution

a First, work out the sum you need to answer for the question. (You cut pieces of rope 85 mm and the other 127 mm, so the sum is $85 + 127$). Work out the calculation (set this out as a **column addition**):

```
      8 5
  + 1 2 7
    2 1 2
    1 1
```

Finally, give your answer in the context of the question (212 mm of rope).

b First, work out the sum you need to answer the question. (You cut 212 mm of rope from 280 mm of rope, you have to take the smaller number from the length of rope you have so the sum is $280 - 212$). Then work out the calculation (set this out as a **column subtraction**):

```
    2 8 0
  - 2 1 2
      6 8
```

Remember you have to subtract the bottom number, 2, from the top number, 0. You cannot take 2 from 0, so you may wish to think of 280 as $200 + 70 + 10$. ←

```
    2 ⁷8 ¹0
  - 2 1 2
      6 8
```

Finally, give your answer in the context of the question (68 mm of rope).

Alternatively, you could say 2 from 0 doesn't go so I will change 8 tens to 7 tens and 10 units. Then $10 - 8 = 2$, $7 - 1 = 6$ and $2 - 2 = 0$. Note we do not need to write this 0 as it is the hundreds column. If a zero would be the first number on the left-hand side of a whole number (value 1 or above) you do not need to put the leading zero.

Try this one. Choose the method you prefer.
$4233 - 107$

Did you get the answer 4126? Well done!

Multiplication and division

Look at this number line.

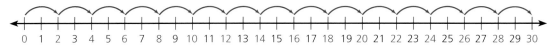

You should already know the multiplication table for 2 and so be able to double numbers.

You can use your knowledge of this table to build up other multiplication tables. It may help to draw your own table as you build up your understanding of the 3, 4, 5 and 10 times tables.

×	1	2	3	4	5	6	7	8	9	10
2	2	4	6	8	10	12	14	16	18	20

The multiplication table for 4 is twice as much as the table for 2.

×	1	2	3	4	5	6	7	8	9	10
2	2	4	6	8	10	12	14	16	18	20
4	4	8	12	16	20	24	28	32	36	40

You can see this on the number line where the leaps are in 4s.

You can build up your table for 3s by looking at what happens when you move along the number line in jumps of 3.

So the multiplication table for 3s looks like this:

×	1	2	3	4	5	6	7	8	9	10
3	3	6	9	12	15	18	21	24	27	30

Multiplying by 10 makes each number ten times as big. In moving the digit from the units to the tens, the space is filled with a zero, so it is easy to remember.

×	1	2	3	4	5	6	7	8	9	10
10	10	20	30	40	50	60	70	80	90	100
5										

We complete this table for the number 5 by halving each number in the × 10 row. It is very useful to know how to multiply by 5, so memorise this table.

×	1	2	3	4	5	6	7	8	9	10
10	10	20	30	40	50	60	70	80	90	100
5	5	10	15	20	25	30	35	40	45	50

Activity

Try to learn the times tables for 2, 3, 4, 5 and 10.

Some people are confident when saying 2, 4, 6, 8, 10, 12, 14, 16, 18, 20 but cannot say quickly what 6×2 equals. Therefore, it is a good idea to learn your times tables by saying 1×2 is 2, 2×2 is 4 (or two 2s are 4), 3×2 is 6 (or three 2s are 6) and so on.

2 × table	3 × table	4 × table	5 × table	10 × table
$1 \times 2 = 2$	$1 \times 3 = 3$	$1 \times 4 = 4$	$1 \times 5 = 5$	$1 \times 10 = 10$
$2 \times 2 = 4$	$2 \times 3 = 6$	$2 \times 4 = 8$	$2 \times 5 = 10$	$2 \times 10 = 20$
$3 \times 2 = 6$	$3 \times 3 = 9$	$3 \times 4 = 12$	$3 \times 5 = 15$	$3 \times 10 = 30$
$4 \times 2 = 8$	$4 \times 3 = 12$	$4 \times 4 = 16$	$4 \times 5 = 20$	$4 \times 10 = 40$
$5 \times 2 = 10$	$5 \times 3 = 15$	$5 \times 4 = 20$	$5 \times 5 = 25$	$5 \times 10 = 50$
$6 \times 2 = 12$	$6 \times 3 = 18$	$6 \times 4 = 24$	$6 \times 5 = 30$	$6 \times 10 = 60$
$7 \times 2 = 14$	$7 \times 3 = 21$	$7 \times 4 = 28$	$7 \times 5 = 35$	$7 \times 10 = 70$
$8 \times 2 = 16$	$8 \times 3 = 24$	$8 \times 4 = 32$	$8 \times 5 = 40$	$8 \times 10 = 80$
$9 \times 2 = 18$	$9 \times 3 = 27$	$9 \times 4 = 36$	$9 \times 5 = 45$	$9 \times 10 = 90$
$10 \times 2 = 20$	$10 \times 3 = 30$	$10 \times 4 = 40$	$10 \times 5 = 50$	$10 \times 10 = 100$

The times table rows for 6, 7, 8 and 9 can be more difficult to learn. You can work some of them out by remembering 6×1 is the same as 1×6 or you can use repeated addition.

6 × table	7 × table	8 × table	9 × table
$1 \times 6 = 6$	$1 \times 7 = 7$	$1 \times 8 = 8$	$1 \times 9 = 9$
$2 \times 6 = 12$	$2 \times 7 = 14$	$2 \times 8 = 16$	$2 \times 9 = 18$
$3 \times 6 = 18$	$3 \times 7 = 21$	$3 \times 8 = 24$	$3 \times 9 = 27$
$4 \times 6 = 24$	$4 \times 7 = 28$	$4 \times 8 = 32$	$4 \times 9 = 36$
$5 \times 6 = 30$	$5 \times 7 = 35$	$5 \times 8 = 40$	$5 \times 9 = 45$
$6 \times 6 = 36$	$6 \times 7 = 42$	$6 \times 8 = 48$	$6 \times 9 = 54$
$7 \times 6 = 42$	$7 \times 7 = 49$	$7 \times 8 = 56$	$7 \times 9 = 63$
$8 \times 6 = 48$	$8 \times 7 = 56$	$8 \times 8 = 64$	$8 \times 9 = 72$
$9 \times 6 = 54$	$9 \times 7 = 63$	$9 \times 8 = 72$	$9 \times 9 = 81$
$10 \times 6 = 60$	$10 \times 7 = 70$	$10 \times 8 = 80$	$10 \times 9 = 90$

Activity

Can you see a pattern with the numbers of the $9 \times$ table?

If you add the two digits together the numbers always add up to 9 (e.g. 18, $1 + 8 = 9$).

Now you have seen all the multiplication tables up to 10, complete your own table with all the multiples in and try to learn them all. Some values have been shown for you in this table.

×	1	2	3	4	5	6	7	8	9	10
1	1	2						8		
2		4		8		12				20
3	3				15				27	
4			12				28			
5				20						50
6		12					42			
7	7					42				
8					40			64		
9			27				63		81	
10	10									100

> ### Learner tip
> Remember that 3×1 is the same as 1×3. If you can't remember what 7×4 is, you may remember 4×7.

Multiplying

There are a number of different methods for multiplying. You may favour one or the other, but you may find it useful to be able to use several of them.

Method 1

For 53×4:

```
    5 3
  ×   4
  ─────
    1 2   (3 × 4)
  2 0 0   (50 × 4)
  ─────
  2 1 2
```

or

```
    5 3
  ×   4
  ─────
  2 0 0   (50 × 4)
    1 2   (3 × 4)
  ─────
  2 1 2
```

It does not matter if you start at the right-hand side and multiply the digit column first or start at the left-hand side and multiply the tens column first (or hundreds if you have any). Just remember what you are multiplying and if it is by ten you need to put the zero in for the place value.

Method 2

Using a grid for 53 × 4:

×	50	3
4	200	12

200 + 12 = 212

Method 3

Using the lattice method for 53 × 4:

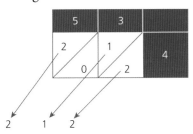

2   1   2

Example

Work out the following:

a 32 × 3

b 18 × 7

Solution

a   3 2

  ×   3

    9 6

You multiply first the units and then the tens by 3.

2 × 3 = 6 and 3 × 3 = 9

b   1 8

  ×   7

    1 2 6

         5

8 × 7 = 56

You write the 6 in the units column and carry 5 tens over.

1 × 7 = 7; 7 + 5 carried over = 12.

These methods will be very useful when you get to Stage 2.

Division

There are two main methods for division: long division and chunking. We will look at both methods here.

Method 1: Long division

For $72 \div 3$:

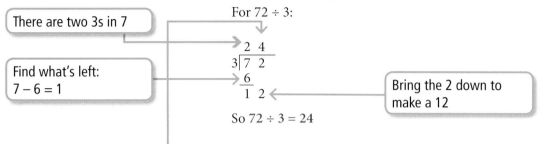

There are two 3s in 7

Find what's left:
$7 - 6 = 1$

There are four 3s in 12

Bring the 2 down to make a 12

$$
\begin{array}{r}
2\ 4 \\
3\overline{)7\ 2} \\
\underline{6} \\
1\ 2
\end{array}
$$

So $72 \div 3 = 24$

Method 2: Chunking

Using this method, you keep taking chunks out that you already know and then work out how many chunks you've taken.

For $72 \div 3$

$$
\begin{array}{rl}
7\ 2 & \\
\underline{-3\ 0} & (10 \times 3) \\
4\ 2 & \\
\underline{-3\ 0} & (10 \times 3) \\
1\ 2 & \\
\underline{-1\ 2} & (4 \times 3)
\end{array}
$$

There are $10 + 10 + 4 = 24$ chunks, so $72 \div 3 = 24$.

Let's work through some examples. You can use either of the methods, but the solution given just shows the method of long division.

Tip for assessment

A common error is not getting the 'order' right when dividing numbers. For example, if the sum is $65 \div 5$, make sure you write $5\overline{)65}$. Use the tasks and tests which follow to make sure you can do this.

Example

Work out the following.

a $96 \div 3$

b $126 \div 7$

Solution

a
$$
\begin{array}{r}
3\ 2 \\
3\overline{)9\ 6}
\end{array}
$$

You divide first the tens and then the units by 3.
$9 \div 3 = 3$ and $6 \div 3 = 2$.

b
$$
\begin{array}{r}
1\ \ 8 \\
7\overline{)1\ 2\ ^5 6}
\end{array}
$$

7 into 1 does not go so you look at the next digit.
7 into 12 is 1 remainder 5.
Look at the remainder together with the next digit.
7 into 56 is 8.

Tasks

1 46 + 32 = 78
   What other calculation can be made using the three numbers 32, 46 and 78?

2 38 + 126 = 164
   What other calculation can be made using the three numbers 38, 126 and 164?

3 Write down the problem in a real context for each of the calculations you wrote down in question 1.

4 Without using a calculator, make as many different number calculations as you can. Each calculation should use three number cards, + or – and =.

2	7	5	11	6	18	13	20

   You could set yourself a challenge by setting a time limit of five minutes to think of and write down the calculations. (12 different calculations is good, 14 is excellent.)

5 Work out the following.
   a 7 + 12
   b 15 – 6
   c 36 + 29
   d 74 – 26
   e 207 + 49
   f 239 – 160

6 Work out the answers to these multiplications.
   a 14×3
   b 23×4
   c 39×5
   d 45×9
   e 124×6

7 Work out the answers to these divisions.
   a $15 \div 3$
   b $\dfrac{12}{4}$
   c $16 \div 4$
   d $20 \div 5$
   e $4\overline{)124}$

Test your knowledge

1 What is 6 + 10 + 9?

   a 16

   b 25

   c 26

   d 160

2 What is the total of 1638 and 519?

   a 1147

   b 2147

   c 2157

   d 6828

3 What is 50 − 36?

   a 14

   b 16

   c 24

   d 26

4 A factory has 280 workers. 34 are away ill.
How many workers are not away ill?

   a 246

   b 254

   c 256

   d 264

5 $80 \times 8 =$

   a 64

   b 640

   c 648

   d 864

6 A person sleeps 8 hours a day.
How many hours will that person sleep in a year (365 days)?

   a 2480

   b 2880

   c 2920

   d 4504

7 $60 \div 5 =$

   a 10

   b 12

   c 15

   d 30

8 $3\overline{)246}$

   a 28

   b 82

   c 208

   d 802

9 $409 \times 7 =$

   a 283

   b 343

   c 2807

   d 2863

10 Share $565 evenly between 5 people. How many dollars does each person receive?

   a $100

   b $103

   c $111

   d $113

Unit 106
Operations on decimal fractions involving tenths and hundredths

Introduction

Mathematics has three ways to describe parts of numbers: using a **fraction**, a **decimal** or a **percentage**. This unit will cover addition, subtraction, multiplication and division of decimal fractions.

It might be useful to cover Unit 101 before tackling this topic, as it provides an introduction to what a fraction means and how decimal place values work, something that is particularly important when learning about money.

This unit also builds on the skills of addition, subtraction, multiplication and division you covered in Unit 105. As you cover new topics in mathematics you will find the ideas often build on something you have covered before, so make sure you understand each topic and ask for help if you are unsure.

You will also find decimal fractions useful when working with measures in the decimal system. Kilograms (for weight), metres (for distance and length) and litres (for capacity) are all based on tens.

Learning objectives

In this unit you will find information on:

- addition, subtraction, multiplication and division of decimal fractions
- measures in the decimal system.

Addition of decimal fractions

Remember the Hundreds, Tens and Units grid we looked at in Numbers and the number system. Each unit is made up of smaller parts called tenths and hundredths.

Hundreds	Tens	Units	.	Tenths	Hundredths
900	90	9		9	90
800	80	8		8	80
700	70	7		7	70
600	60	6		6	60
500	50	5		5	50
400	40	4		4	40
300	30	3		3	30
200	20	2		2	20
100	10	1		1	10

> **Learner tip**
>
> Note that 3.2 is greater than 3.06 as the 2 represents $\frac{2}{10}$ and the 6 represents $\frac{6}{100}$.

A decimal point is used to show where the unit is. The first number after the decimal point represents tenths, the next represents hundredths, and so on. You will be familiar with this from using money.

The dollar is made up of 100 cents so half a dollar is 50 cents. In tenths and hundredths, 50 cents is 5 tenths of a dollar and 0 hundredths.

We can write this as $0.50. There are no dollars so nothing from the hundreds, tens or units columns of the grid and we just need one zero to show this.

This next grid is the same for measuring length: hundreds, tens, units, tenths and hundredths but this time the unit column is metres and the tenths are 10 centimetres.

Hundreds	Tens	Units	Tenths	Hundredths
100 metres	10 metres	metre	10 centimetres	centimetre

Remember **1 metre is the same as 100 centimetres.**

Example 1

Daryl wants to measure his height.

The pointer on this scale shows his height.

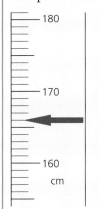

What is his height?

Solution

The pointer is between 160 cm and 170 cm.

There are 10 divisions between 160 cm and 170 cm.

Those 10 divisions cover 10 cm, so 1 division is 1 cm.

The pointer is at the 6th division after 160 cm, so the height is **166 cm.**

Daryl is 166 cm or **1.66 m** tall.

Example 2

Add 2.6 + 1.3

Solution

Line up the columns so the decimal points (.) are on top of each other. This means you are adding the correct units together.

$$\begin{array}{r} 2.6 \\ +1.3 \\ \hline 3.9 \end{array}$$ 6 + 3 = 9, 2 + 1 = 3

the answer is **3.9**

This could be 3.9 metres or 3.9 kilograms but there are no units in the question so we just give the answer as 3.9

Remember to line up the columns so the decimal points (.) are on top of each other.

Example 3

Add 14. 5 and 1.86

Solution

$$\begin{array}{r} 14.5 \\ + 1.86 \\ \hline 16.36 \\ {\scriptstyle 1} \end{array}$$

There is no other number in the hundredths column so 0 + 6 hundredths = 6 hundredths, 5 tenths + 8 tenths = 13 tenths which is 1 unit and 3 tenths, so carry 1 across into the units column and write 3 tenths. 4 + 1 + 1 = 6 units. There is only one figure in the tens column so 1 + 0 = 1.

The answer is **16.36**

Activity

Try these.

a 13.6 + 2.3

b 127.5 + 15.6

c 3.25 m + 1.7 m

d $15 + $2.50

Remember 15 is a whole number so it can be written as 15.0

Learner tip

You need to be really confident with knowing what place values in numbers mean, so for example for 348.7 you need to know that the 3 stands for 3 hundreds (300), 4 for 4 tens (40), 8 for 8 units and 7 for 7 tenths (0.7). Mistakes are easily made by not lining up the correct place values when you are under pressure, 300 + 1.35 is 301.35.

Remember that you can add a zero before or after a decimal number: 300 is the same as 300.00 and 2.5 is the same as 02.50. Think about your hundreds, tens, units, tenths and hundredths columns to check.

You can't add a zero to the right of a whole number unless you insert the decimal point: 20 is very different from 200 but 20 is the same as 20.0 or even 20.00

Subtraction of decimal fractions

Subtraction of decimal numbers is completed in the same way as subtraction of hundreds, tens and units. Remember to line up the columns so the decimal points (.) are on top of each other.

9.6 − 5.3

Line up the columns so the decimal points (.) are on top of each other. This means you are subtracting the correct units.

$$\begin{array}{r} 9\,.\,6 \\ -\,5\,.\,3 \\ \hline 4\,.\,3 \end{array}$$ 6 − 3 = 3, 9 − 5 = 4

the answer is **4.3**

This could be 4.3 litres or 4.3 kg but there are no units in the question so we just give the answer as 4.3

26.5 − 3.47

$$\begin{array}{r} 2\,6\,.\,4\,\overset{10}{\cancel{5}} \\ -\,\;\;3\,.\,4\,7 \\ \hline 2\,3\,.\,0\,3 \end{array}$$

The answer is **23.03**

$30 − $6.50

$$\begin{array}{r} 2\overset{9}{\cancel{3}}\,\overset{10}{\cancel{0}}\,.\,\overset{10}{0}\,0 \\ -\;\;\;\;6\,.\,5\,0 \\ \hline 2\;\;3\,.\,5\,0 \end{array}$$

The answer is **$23.50**

> ## Activity
> Try these:
> **a** 137.6 − 6.3
> **b** 27.5 − 15.6
> **c** 3.25 m − 1.5 m
> **d** $145 − $32.50

> Remember to line up the columns so the decimal points (.) are on top of each other.

> 26.5 can also be written as 26.50

> In the hundredths column, you can't take 7 away from nothing so you need to change one tenth into 10 hundredths. 10 − 7 hundredths = 3 hundredths.
>
> We started with 5 tenths but took one to make hundredths so now there are only 4 tenths, 4 − 4 = 0 tenths.
>
> Then subtract in the units column: 6 − 3 = 3 units. There is only one figure in the tens column so 2 − 0 = 2.

> 0 − 0 = 0. You can't take 5 from 0, so change 1 unit into 10 tenths but there are no units so change 1 ten to 10 units first. Then 10 − 5 = 5. You only have 9 units now, so 9 − 6 = 3 and 2 tens − 0 = 2

Multiplication of decimal fractions

It is important to be able to recall quickly the multiplication tables up to 10×10 (see Unit 105). When these have been mastered, you will have the confidence to move onto multiplying and dividing decimals.

Multiplying by 10 and 100

Here are two entries in the 10 times table.

$5 \times 10 = 50 \qquad 10 \times 10 = 100$

You can see that to multiply by 10 you move the units into the tens column, the tens into the hundreds column and so on. You put a zero in the units column.

So, for example,

$25 \times 10 = 250, 564 \times 10 = 5640, 120 \times 10 = 1200$

In the same way,

$4 \times 100 = 400, 6 \times 100 = 600$

You can see that to multiply by 100 you move the units into the hundreds column, the tens into the thousands column and so on.

You put zeros in the units and the tens columns.

> ### Example
> Write down the answers to these.
> **a** 56×10
> **b** 47×100
> **c** 129×10
> **d** 420×100
>
> ### Solution
> **a** 560
> **b** 4700
> **c** 1290
> **d** 42000

Now that you know how to multiply integers (whole numbers) by 10 and 100, you can use place value tables to help you multiply decimals by 10 and 100. Let's use a worked example to see how to do this.

Example

Work out these multiplication sums.

a 35.6×10

b 35.6×100

c 7.8×10

d 7.8×100

Solution

a To multiply by 10 you move each digit one place to the left.

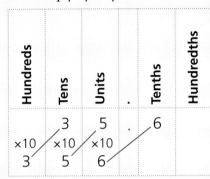

$35.6 \times 10 = 356$

b To multiply by 100 you move each digit two places to the left. Notice that you have to put a zero in the units column otherwise the number would read 356 rather than 3560.

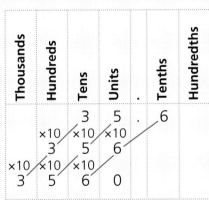

$35.6 \times 100 = 3560$

c To multiply by 10 you move each digit one place to the left.

$7.8 \times 10 = 78$

d To multiply by 100 you move each digit two places to the left. Notice that you have to put a zero in the units column otherwise the number would read 78 rather than 780.

$7.8 \times 100 = 780$

Activity

In pairs, create and answer multiplications ($\times 10$ and $\times 100$) for whole numbers and for decimal fractions.

Multiplying a decimal fraction by an integer (a whole number)

Compare these two multiplications for finding the cost of three items at $4.95 each.

Working in cents

You know that $4.95 = 495 cents.

$$
\begin{array}{r}
4\,9\,5 \\
\times \quad 3 \\
\hline
1\,4\,8\,5 \quad = \$14.85
\end{array}
$$

Multiply first the units, then the tens and then the hundreds by 3.

Write the units under the units, the tens under the tens, and so on.

Convert your answer from cents to dollars by dividing by 100.

1485 = $14.85

Working in dollars

$$
\begin{array}{r}
4.9\,5 \\
\times \quad 3 \\
\hline
\$1\,4.8\,5
\end{array}
$$

To multiply a decimal by an integer, put the decimal points under each other.

Make sure you line up your work carefully.

Put the first digit you work out under the last decimal place.

In each case, the digits are the same.

Multiply each number by 4, starting with the decimal number (2 tenths)

Make sure you line up your work carefully.

Multiply each number by 5, starting with the 1 hundredth then 5 tenths

Make sure you line up your work carefully.

Example

Let's try some worked examples.

a 13.2 × 4
b 20.51 × 5

Solution

a
$$
\begin{array}{r}
1\,3.2 \\
\times \quad 4 \\
\hline
5\,2.8
\end{array}
$$

b
$$
\begin{array}{r}
2\,0.5\,1 \\
\times \qquad 5 \\
\hline
1\,0\,2.5\,5 \\
{\scriptstyle 1 \quad 2}
\end{array}
$$

$0.01 \times 5 = 0.05$; $0.5 \times 5 = 2.5$, carry 2 into the units column; $0 \times 5 = 0$ but $0 + 2 = 2$. $2 \times 5 = 10$, put 0 in tens column and carry 1 into the hundreds column.

Activity

Try these.

a 7.8×6

b 12.9×3

c 123.5×2

d 66.8×4

Tips for assessment

To check you have the decimal point in the correct place in your answer, you can estimate your answer. 20.5×5 is going to be more than 10 and nearer 100 than 1000.

Division of decimal fractions

You should understand that multiplication and division are opposite processes. So if you know that $4 \times 6 = 24$, you should also know that $24 \div 4 = 6$ and $24 \div 6 = 4$.

Dividing by 10 and 100

Since dividing is the reverse of multiplying, to divide by 10 you move the digits one place to the right and take a zero off.

To divide by 100 you move the digits two places to the right and take two zeros off.

Now that you know how to multiply and divide integers by 10 and 100, you can use place value tables to help you divide decimals by 10 and 100. Let's use a worked example to see how to do this.

Example

a $580 \div 10$

b $1400 \div 100$

c $36\,000 \div 100$

d $60\,000 \div 100$

Solution

a $580 \rightarrow 58$

b $1400 \rightarrow 14$

c $36000 \rightarrow 360$

d $60000 \rightarrow 600$

Example

Work out this division sum.

a $435.2 \div 10$

Solution

a To divide by 10 you move each digit one place to the right.

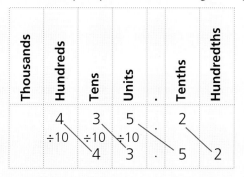

$435.2 \div 10 = 43.52$

Examples

Work out these sums.

b 435.2 ÷ 100

c 79.2 ÷ 100

d 79.2 ÷ 10

Solutions

b To divide by 100 you move each digit two places to the right.

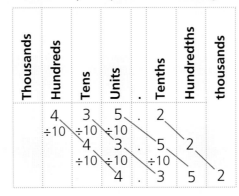

435.2 ÷ 100 = 4.352

c To divide by 100 you move each digit two places to the right. You usually write a zero when there are no digits in front of the decimal point. This makes the number easier to read.

79.2 ÷ 100 = 0.792

d 79.2 ÷ 10 = 7.92

Activity

In pairs, create and answer division sums (÷10 and ÷ 100) for whole numbers and for decimal fractions.

Dividing a decimal fraction by an integer (a whole number)

Working in cents

You know that $32.60 ÷ 4 is 3260 cents shared by four.

For 3260 ÷ 4 this is just one method you can use. See Unit 105 on multiplication and division of whole numbers.

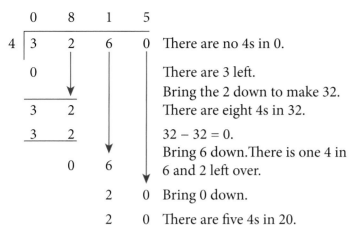

So 3260 ÷ 4 = 815

Convert your answer from cents to dollars by dividing by 100.

815 ÷ 100 = 8 dollars and 15 cents

Working in dollars

$$\begin{array}{r} 8.15 \\ 4\overline{)3\ 2.6\ 0} \end{array}$$

Complete the calculation as before and line up the decimal point on the answer line above the decimal point in the number before you start to divide.

Let's try some worked examples.

Example

a 72.6 ÷ 3

b 145.22 ÷ 2

Solution

a

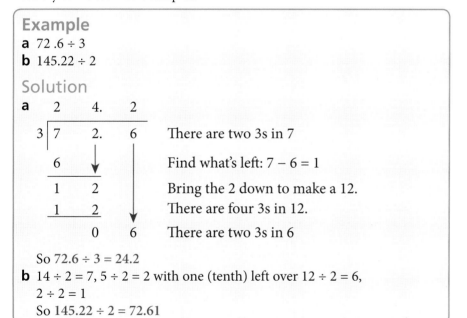

So 72.6 ÷ 3 = 24.2

b 14 ÷ 2 = 7, 5 ÷ 2 = 2 with one (tenth) left over 12 ÷ 2 = 6, 2 ÷ 2 = 1

So 145.22 ÷ 2 = 72.61

Tips for assessment

Remember 21 ÷ 7, $7\overline{)21}$ and $\frac{21}{7}$ all mean 21 divided by 7.

To check you have the decimal point in the correct place in your answer, you can estimate your answer. 21 ÷ 7 is going to be less than 10.

Activity

Try these.

a 46.8 ÷ 6

b 38.7 ÷ 3

c 247 ÷ 2

d 267.2 ÷ 4

Tasks

Now have a go at these tasks.

1 Work out these.

   a   6.7 2
      +   7.1 9

   b   1 8.9 5
      + 2 3.1 4

   c   2 7.5 4
      + 8 3.6 1

   d   1 6.7 4
      + 4 3.9 7

2 Work out these.

   a   1 6.7 8
      −   7.1 3

   b   2 8.7 5
      − 1 3.8 4

   c   1 2 8.3 6
      −   7 3.5 2

   d   4 3 9.8 7
      − 2 1 8.0 3

3 Work out these.

   a 6.85 m + 0.4 m

   b $16.83 + 94 cents

   c 12 litres + 0.5 litre

   d 3.6 kg + 2.7 kg

4 Find the weight of five boxes at 3.5 kg each.

5 In the long jump, Janet jumps 13.42 m and Delia jumps 15.18 m. Find the difference between the lengths of their jumps.

6 The times for the first and last places in a 200-metre race were 24.42 seconds and 27.38 seconds. Find the difference between these times.

7 Work out these.

   a 2.10 + 3.45

   b 5.78 + 2.82

   c 7.15 − 6.13

   d 123.67 − 65.77

8 Work out these.

   a 4.37 + 6.53

   b 15.78 + 9.89

   c 6.18 + 6.32

   d 15.42 − 9.34

9 Work out these.

    a 1.6 + 3.4

    b 4.9 + 5.21

    c 21.99 − 11.9

    d 5.53 − 3.09

10 Work out these.

    a 4.9 + 5.6

    b 34.94 + 7.62

    c 8.6 − 3.4

    d 9.42 − 8.57

11 Multiply each of these numbers by 10.

    a 56

    b 7.9

    c 7.34

    d 8.3

12 Multiply each of these numbers by 100.

    a 56

    b 7.9

    c 7.34

    d 8.3

13 Divide each of these numbers by 10.

    a 11.8

    b 276

    c 30.06

    d 28.39

14 Divide each of these numbers by 100.

    a 11.8

    b 8.3

    c 276

    d 30.06

15 Work out these.

    a 4.8×4

    b 10.3×22

    c 2.4×63

    d 2.12×8

16 Work out these.

    a $13.6 \div 4$

    b $11.4 \div 3$

    c $21.6 \div 6$

    d $58.1 \div 7$

1   5.9 1
    + 8.7 2

  a 1.93

  b 11.12

  c 13.63

  d 14.63

2   3 3.5 1
    + 9.8 6

  a 32.37

  b 43.27

  c 43.37

  d 132.11

3   1 3.4 9
    − 5.1 8

  a 8.21

  b 8.31

  c 18.67

  d 81.69

4 47.51 − 26.74

  a 20.77

  b 21.23

  c 21.77

  d 21.87

5 34.02 − 14.89

  a 19.13

  b 19.23

  c 20.23

  d 20.87

6 19.21 − 17.04

  a 2.17

  b 2.23

  c 2.25

  d 2.27

7 5.9×10

  a 0.59

  b 5.09

  c 59

  d 509

8 9.62×100

  a 0.962

  b 96.2

  c 962

  d 9620

9 364 divided by 100

  a 0.364

  b 3.064

  c 3.64

  d 36.4

10 $9.68 \div 4$

  a 2.12

  b 2.17

  c 2.32

  d 2.42

Unit 107
Operations on common fractions involving halves and quarters

Introduction

A fraction is a way of describing a number which is less than one whole. In Unit 101 we introduced the simple fractions:

- one-half $\left(\dfrac{1}{2}\right)$

- one-quarter $\left(\dfrac{1}{4}\right)$.

In this unit we develop this knowledge.

Learning objectives

In this unit, you will find information on:

- adding and subtracting these simple fractions.

What a fraction means

A fraction is a way of describing a number which is less than one whole. Think of a pizza.

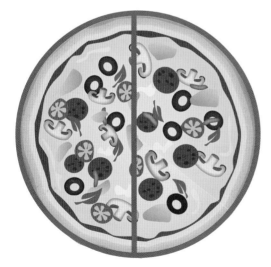

Here it is sliced into 2 equal pieces or **halves**. As a fraction of the whole pizza, we write that one slice as $\dfrac{1}{2}$.

This is said as 'one-half'.

In a fraction, the top number is called the **numerator** and tells you how many parts the fraction shows. The bottom number is called the **denominator** and tells you how many equal parts the whole is split into. So if the pizza is split into four equal pieces, one piece is $\frac{1}{4}$. This is said as 'one-quarter'.

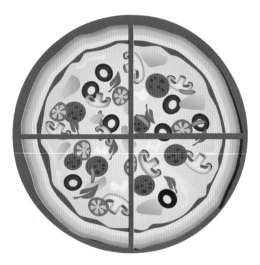

Notice that the size of the slice for one-quarter of a pizza is smaller than for one-half. You can see that two of these one-quarter slices would give you the same amount of pizza to eat as having one-half of the first pizza.

Activity

Trace each shape on a sheet of paper then colour in half of each of these shapes, like this:

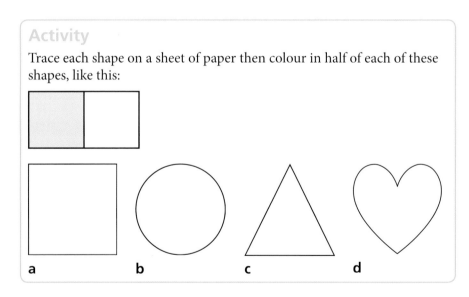

Activity

Trace these shapes on a sheet of paper then colour in one-quarter of each of these shapes like this

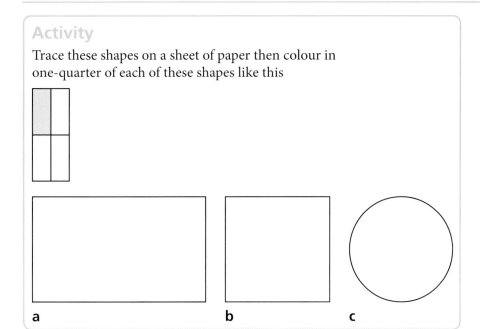

a **b** **c**

Some shapes can divide into 4 equal pieces in different ways:

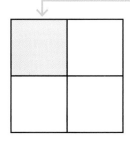 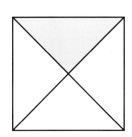 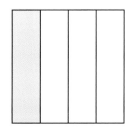

Remember it is written $\frac{1}{4}$ to show the whole is divided into 4 and we have shaded in 1 of these.

What would the fraction $\frac{2}{4}$ look like as a pizza picture? This would be a pizza cut into four equal slices (the bottom number or denominator) and you taking two of these slices (the top number or numerator) to eat.

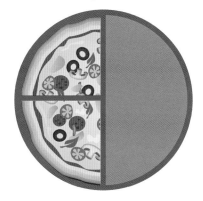

What fraction is this the same as? Yes, this is the same as $\frac{1}{2}$

$$\frac{2}{4} = \frac{1}{2}$$

Activity

What would the fraction $\frac{3}{4}$ look like?

Trace this circle on a sheet of paper then colour in $\frac{3}{4}$ of this circle.

Real world maths

What you have drawn will be like a pizza cut into four equal slices (the bottom number or denominator) and you having three of these slices (the top number or numerator) left to eat.

Or $\frac{3}{4}$ of a bar of chocolate: you eat two pieces and leave six pieces.

Adding fractions

To add fractions, all the fractions must be the same size such as all halves or all quarters.

$$\frac{1}{2} + \frac{1}{2} = \frac{2}{2}$$

Think of the pizza. If you have two halves, you have a whole. So one-half add one-half gives you two-halves, which is the same as a whole.

We can add quarters in the same way:

$$\frac{1}{4} + \frac{1}{4} + \frac{1}{4} + \frac{1}{4} = \frac{4}{4}$$

Think of the pizza again. If you have four-quarters, you have a whole. So one-quarter add one-quarter add one-quarter add one-quarter gives you four-quarters, which is the same as a whole.

How can we add one-half and one-quarter?

We need to change the half to quarters. Think of the pizza or a piece of paper.

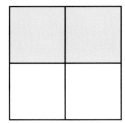

We can see that a whole is 4 quarters or 2 halves. We can also see that $\frac{1}{2}$ is the same as two-quarters or $\frac{2}{4}$.

If we add two-quarters and one-quarter, we get three-quarters.

$$\frac{2}{4} + \frac{1}{4} = \frac{3}{4}$$

We can add these across the numerators (the top number) because this tells us how many parts we have. But we can only do this if the denominators (the bottom number) are the same. Note we do not add the bottom numbers.

We can also write this as $\frac{2+1=3}{4}$

What is $\frac{3}{4} + \frac{1}{4}$?

$\frac{3+1=4}{4} = \frac{4}{4}$ is one whole so $\frac{3}{4} + \frac{1}{4} = 1$

Subtracting fractions

To subtract fractions all the fractions must be the same size such as all halves or all quarters.

$$\frac{3}{4} - \frac{1}{4} = \frac{2}{4}$$

Think of the pizza. If there are three-quarters on a plate and you take one-quarter, there are two-quarters left on the plate.

> We can say there are two-quarters left but we usually call two-quarters one-half. Draw a pizza to check this if you are not certain.

What is $\frac{1}{2} - \frac{1}{4}$?

To subtract fractions all the fractions must be the same size. A quarter is not as big as a half so we need to make the half into quarters. $\frac{1}{2} = \frac{2}{4}$

Now we can do the subtraction sum: $\frac{2}{4} - \frac{1}{4} = \frac{1}{4}$

Think of the pizza again. If you had a pizza cut into 4 and half of it was on the plate, there would be two-quarters on the plate. You take away one-quarter and one-quarter is left.

You often need to subtract fractions from a whole.

You have 3 litres of oil and use half a litre. How much oil is left?

Think of this as pizzas if it helps. You can change 3 into 6 halves.

$$\frac{6}{2} - \frac{1}{2} = \frac{5}{2}$$

Think of the pizzas. How many whole and half pizzas are left?

2 whole and one-half pizzas = $2\frac{1}{2}$

Therefore, there are $2\frac{1}{2}$ litres of oil left.

Alternatively, you can just change one litre into 2 halves and remember you have 2 full litres. This is a similar approach to what you did with hundreds, tens and units in Unit 101.

> Use whichever method you find easier.

$$\frac{2}{2} - \frac{1}{2} = \frac{1}{2} + 2 \text{ whole litres} = 2\frac{1}{2} \text{ litres.}$$

Fractions of quantities

Fractions don't just apply to slices of a whole. If you have 6 biscuits and eat one-half of them, how many have you eaten? This is the same as making two equal piles of biscuits (remember the bottom number, or denominator, is dividing into two here). The top number tells us how many parts or piles, so here it is one pile of three biscuits.

$\frac{1}{2}$ of 6 = 3

If you have 10 sweets and give your friend half of them, how many would your friend have? Make two equal piles (remember the bottom number or denominator is dividing into two here) and give your friend half.

$\frac{1}{2}$ of 10 = 5

> ### Activity
>
> Try finding half of different even numbers. You can use counters or pencils to help you.

What happens if you have an odd number, for example 5? You can make two equal piles of 2 but you have one left over. If this is a biscuit you could break this in half and then have two equal piles with $2\frac{1}{2}$ biscuits in each pile. Half of 5 is $2\frac{1}{2}$.

If you want to find half of a pile of 7 sweets, you can make 2 piles of 3 with one left over.

Half of 7 is $3\frac{1}{2}$.

To calculate one-half of an amount, you can either multiply the amount by $\frac{1}{2}$ or divide it by 2. You will get the same answer in both cases.

Here you have 24 cans. Using our two ways of calculating half of 24, we find that the answer is 12.

$24 \times \dfrac{1}{2} = 12$

$24 \div 2 = 12$

So what happens if you have 40 apples and want to work out one-quarter?

One-quarter of 40 is either

$40 \times \dfrac{1}{4} = 10$

$40 \div 4 = 10$

> **Tips for assessment**
>
> In a fraction, the denominator (bottom number) tells you how many you have split the whole into and the numerator (top number) tells you how many of those you need.
>
> An image of a pizza sometimes helps you to imagine what a fraction looks like.
>
> When finding a unit fraction (a fraction where the numerator is 1) of an amount, you can either multiply by the fraction or divide by the denominator.

Tasks

1 Copy and complete the pictures, shading in the fraction shown.

   a Show $\frac{1}{2}$ **b** Show $\frac{1}{4}$

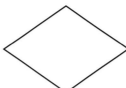

   c Show $\frac{3}{4}$

2 Write down the fraction shown in each case.

   a   **b**

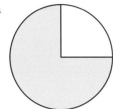 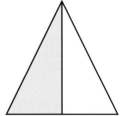

3 $\frac{1}{2} + \frac{1}{4} =$

4 $2\frac{1}{2} + \frac{1}{4} =$

5 $1\frac{1}{2} + \frac{3}{4} =$

6 $\frac{3}{4} - \frac{1}{4} =$

7 $2 - \frac{1}{4} =$

8 $1\frac{3}{4} - \frac{1}{2} =$

9 Find $\frac{1}{2}$ of these quantities.

   a 16

   b 50

10 Find $\frac{1}{4}$ of these quantities.

   a 8

   b 80

Test your knowledge

1 What fraction of the shape is shaded?

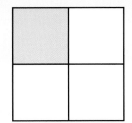

  a $\frac{1}{4}$

  b $\frac{1}{2}$

  c $\frac{1}{3}$

  d $\frac{3}{4}$

2 $\frac{1}{2} + \frac{1}{4} =$

  a $\frac{2}{6}$

  b 1

  c $\frac{3}{4}$

  d $\frac{2}{8}$

3 $1\frac{1}{2} - \frac{1}{4} =$

  a $1\frac{1}{4}$

  b $1\frac{1}{2}$

  c $1\frac{1}{6}$

  d $\frac{3}{4}$

4 There are 2 litres of milk in a carton. Someone uses half a litre. How much milk is left in the carton?

  a $1\frac{1}{2}\,\ell$

  b $\frac{1}{2}\,\ell$

  c $2\frac{1}{2}\,\ell$

  d $1\frac{1}{4}\,\ell$

5 $\frac{3}{4}\text{m} + \frac{1}{2}\text{m} =$

  a $\frac{4}{6}\text{m}$

  b $1\,\text{m}$

  c $1\frac{1}{4}\text{m}$

  d $\frac{4}{8}\text{m}$

6 A dressmaker has 3 m of material. She uses $1\frac{1}{2}\text{m}$. How much material does she have left?

  a $4\frac{1}{2}\text{m}$

  b $2\,\text{m}$

  c $1\frac{1}{2}\text{m}$

  d $1\,\text{m}$

7 Find half of 24.

  a 2   **c** 6

  b 5   **d** 12

8 What is $\frac{1}{2}$ of 100?

  a 5   **c** 25

  b 20   **d** 50

9 Find one-quarter of 40.

  a 4   **c** 10

  b 8   **d** 20

10 What is $\frac{1}{4}$ of 100?

  a 14   **c** 25

  b 20   **d** 50

Unit 108
Appropriate strategies and use of the calculator

Introduction

This unit is about solving mathematical problems in everyday situations, for example:

- working out a total cost to see if you have enough money to pay
- working out if your change is correct.

Learning objectives

At Stage 1, you are required to:

- recognise the operations required to solve a problem
- use checking strategies
- use mathematical terms in everyday conversation.

Appropriate strategies and use of the calculator

These are the five steps to solving a problem.

1 Read the question

It sounds obvious, but it is important.

Many mistakes come from not reading the question carefully.

Look for what kind of answer will be needed.

Answers may be:

- a total cost
- a time or date
- a number, such as a number of people or cars
- a decision, such as 'Does he have enough money to pay?'
- a choice of items to buy for less than a given amount of money.

2 Pick out key information

Pick out key details from the question, such as numbers, costs or times.

You could highlight, underline or copy them.

Look for key words that give a clue to the calculations you'll need to make.

These words mean that you'll probably need to add numbers:

- plus
- and
- with
- total
- altogether.

These words mean that you'll probably need to take away numbers:

- minus
- less
- difference
- without
- How much more...? or How many more ...?
- How much is left over ...? or How many are left over ...?

These words may mean that you'll need to multiply numbers:

- each
- every.

These words may mean that you'll need to divide numbers:

- share
- split.

Tips for assessment

In City & Guilds' exams, you cannot use your calculator, so practise manual methods too.

3 Work it out

Use a calculator or use a manual method.

Whatever method you choose, show your working.

Write down what you put into the calculator and the results you get.

4 Check the number you get from your working

Is the number a sensible answer? For example, if you work out the cost of a cup of tea and a biscuit at $500, there may be something wrong.

You should also check that you have used the correct numbers from the question. It's easy to put 54 in the calculator instead of 45 or write the number down wrong.

Look back at the question to make sure you haven't missed anything.

Check the working by

- reverse calculation – working backwards
- estimation – rounding numbers to give a rough check.

You can find out more about these in Unit 101.

5 Present your final answer correctly

Include the correct units.

Round an answer sensibly.

An amount of money should be given as

- a whole number of dollars, for example $17, or
- dollars and cents to two decimal places, for example $4.50.

Answer the question – it may require a decision or a choice.

To sum up:

1 Read the question
2 Pick out the key information
3 Work it out
4 Check the answer
5 Present the answer

Solving problems

Example 1

Here is some information about Andy's mobile phone contract.

Free every month
250 minutes talk time
500 texts
500 MB data

He has used 178 minutes of talk time so far this month.

How many more minutes are left this month?

Solution

1 Read the question.
   The question asks 'How many more minutes....?'
   The answer is going to be a number of minutes.
2 Pick out the key information
   250 minutes of talk time a month, and 178 minutes used so far this month.
   'How many ... left?' means that we have to take something away.
3 Work it out:
   $250 - 178 = 72$
4 Check the answer.
   72 looks like a sensible answer.
   Rounding to the nearest 10, $250 - 180 = 70$, which is close.
5 Present the answer:
   72 minutes (this is the correct answer with units)

Example 2

Marva saves $72 every month for 6 months.

What is her total saving?

Solution

1 Read the question.

The question asks for total savings.

The answer is going to be an amount of money.

2 Pick out the key information:

$72 a month

6 months

'every month' is a clue that we may have to multiply.

3 Work it out:

$72 \times 6 = 432$

4 Check the answer.

432 looks like a sensible answer.

Rounding to the nearest 10, $70 \times 6 = 420$, which is close.

5 Present the answer:

$432 (this is the correct answer with units)

Order of operations

Does order matter?

Activity

Try working out simple sums to see if the order matters.

Order does not matter for addition problems.

$36 + 6 = 42$

$6 + 36 = 42$

Order does matter for subtraction problems.

If you have $42 and spend $36 the calculation is

$42 - 36 = 6$

$36 - 42$ implies you have $36 and spend $42. You do not have enough money so you owe $6.

Order does not matter for multiplication problems.

$3 \times 6 = 18$

$6 \times 3 = 18$

but note that three lots of 6 is not actually the same as 6 lots of 3 and the context may make it matter. Three people spending $6 gives the same amount of money as six people spending $3 but this may concern you if you are one of the people spending your money.

Order does matter for division problems.

$4 \div 2 = 2$

$2 \div 4 = \dfrac{1}{2}$

Checking strategies

You can check your answer in various ways:

- perform the calculation in a different way, e.g. 30 + 8 can be recalculated as 8 + 30
- a reverse calculation, e.g. 30 + 8 = 38, so check by 38 − 8
- roughly estimate the answer, e.g. 30 + 8 , so work out 30 + 10 and the answer will be around 40
- judge if the answer is sensible, e.g. 30 + 8 is not going to be hundreds
- use a calculator if you have one. ←——————————————

> You will not have a calculator for the City & Guilds exam.

Example

Check these calculations and show your check. Try to use different methods.

a 126 + 32 = 446
b 50 × 10 = 500
c 630 ÷ 3 = 31
d 342 − 68 = 274

Solution

a Check: 130 + 30 = 160 This is not close to 446, so an incorrect answer
b Check: 500 ÷ 10 = 50, so a correct answer
c Check: not a sensible answer for dividing 600 by a small number like 3, so an incorrect answer
d Check: 274 + 68 = 342, correct answer

> The checks shown are examples, they are not the only checks that could be used.

Use mathematical terms

At Stage 1 you should be familiar with these mathematical terms:

- **less than** This means 'lower' or 'not as many', so 6 is less than 10 as it is a lower number. If you have 6 coconuts you do not have as many as someone who has 10 coconuts.

- **equal to** This means 'the same as', so $\frac{1}{2}$ is the same as 0.5. Two-quarters are equal to one-half.

- **greater than** This means 'higher' or 'more than', so 10 is greater than 6 as it is a higher number. If you have 10 coconuts you have more than someone who has 6 coconuts.

- **approximately** This means 'fairly correct' or 'nearly right', so if you estimate an answer, you are not giving the correct answer but one that is close to the correct answer. 1 centimetre is approximately the distance across my little finger. It is not exactly 1 cm but is close enough for me to know how long 1 cm is.

Activity

In pairs, give your partner some calculations to check.

Example

Use each of the mathematical terms once in the statements below.

54 is _____ 50

9×9 is _____ 80

36 is _____ 40

100 centimetres is _____ 1 metre.

Solution

54 is **greater than** 50

9×9 is **approximately** 80

36 is **less than** 40

100 centimetres is **equal to** 1 metre.

Activity

Use each of the mathematical terms to make your own statements.

Tasks

1 Describe the steps you take to solve this problem and decide whether Olivia met her target. Show your calculations.

Olivia sells furniture.

She must make a total number of 500 sales to meet her target.

These are the sales she has made. Did she meet her target?

Item	Number of sales
Bed	247
Sofa	79
Table	162

2 Are these statements correct?

a $36 \times 3 = 3 \times 36$

b $45 \div 9 = 5$

c $\frac{1}{2}$ is less than $\frac{1}{4}$

3 How can you check $67 - 9 = 58$?

Test your knowledge

1 Which statement is correct?

  a 16 is equal to 10

  b 16 is approximately 10

  c 16 is less than 10

  d 16 is more than 10

2 Which statement is correct?

  a A dollar is approximately 100 cents

  b A dollar is equal to 100 cents

  c A dollar is more than 100 cents

  d A dollar is less than 100 cents

3 A customer buys 5 items costing \$7 each. Which calculation should he use to find the total cost?

  a $5 + 7$ **c** 5×7

  b $7 - 5$ **d** $7 \div 5$

4 How can you check $100 \div 5 = 20$?

  a $5 + 20$ **c** $100 - 20$

  b 5×20 **d** $100 + 20$

5 How can you show $\frac{1}{2}$ on the calculator?

  a 1.2   **c** 0.5

  b 0.2   **d** 5

6 Which two will give the same answer?

  A $6 + 6 + 6$ **a** A and B

  B $18 - 6$ **b** B and C

  C 6×3 **c** C and D

  D $18 \div 6$ **d** A and C

STAGE 2

Unit 201
Place value

Introduction

The way our number system works helps us to describe and solve problems. We represent numbers in different ways. Sometimes we write numbers in words (one hundred) and sometimes we write numbers in figures (100). This unit builds on your work at Stage 1, especially the first unit on numbers and the number system (Unit 101) and the unit on decimal fractions (Unit 106).

You will develop your knowledge of place value to include whole numbers up to one million (1 000 000), parts of a whole – decimal fractions to thousandths $\left(\frac{1}{1000}\right)$, and writing these numbers in words and figures.

This will help you to distinguish between numbers of different magnitude, which means you will be able to order numbers by value and identify digits showing different values. For example, you will recognise that 5461.3 has a higher value than 4561.3 and in both numbers the 3 has the value of three-tenths.

Learning objectives

In this unit you will find information on:

- place value including whole numbers up to one million (1 000 000)
- parts of a whole, decimal fractions to thousandths $\left(\frac{1}{1000}\right)$
- writing these numbers in words and figures.

This will help you to prepare for questions about:

- distinguishing between numbers of different magnitude

Place value

This unit will build on skills from your knowledge of how the number system works, for example we can read or write 645 as six hundred and forty-five.

645 means 6 hundreds, 4 tens and 5 units.

What is this number?

256.9

You can see this in the grid on the next page.

> This number reads, two hundred and fifty-six point nine.
>
> It means two hundreds, five tens, six units and nine tenths.

Million	Hundred thousand	Ten thousand	Thousand	Hundred	Tens	Units	.	$\frac{1}{10}$	$\frac{1}{100}$	$\frac{1}{1000}$
				2	5	6	.	9		

Now look at larger numbers. After 999, we have a thousand (1 000).

What is this number?

3 256 ←

This number reads, three thousand, two hundred and fifty-six. There are no numbers after the decimal point so the decimal point is not shown.

It means three thousands, two hundreds, five tens and six units.

Now look at larger numbers. After nine thousand nine hundred and ninety-nine, 9 999, we have ten thousand (10 000).

What is this number?

21 256

This number reads, twenty-one thousand, two hundred and fifty-six. There are no numbers after the decimal point so the decimal point is not shown.

It means two ten-thousands, one thousand, two hundreds, five tens and six units.

Now look at larger numbers. After ninety-nine thousand nine hundred and ninety-nine, 99 999, we have one hundred thousand (100 000).

What is this number?

210 256 ←

This number reads, two hundred and ten thousand, two hundred and fifty-six. There are no numbers after the decimal point so the decimal point is not shown.

It means two hundred-thousands, one ten-thousand, zero thousands, two hundreds, five tens and six units.

Now look at larger numbers. After nine hundred and ninety-nine thousand nine hundred and ninety-nine, 999 999, we have one million (1 000 000).

Activity

Try these.

a What is 39 045 in words?

b What is 100 002 in words?

c What is 100.75 in words?

d What is three thousand and twenty-five in figures?

e What is one million in figures?

f What is thirty-nine point two in figures?

Activity

In pairs, take it in turns to say a number to your partner and ask your partner to write the number in words and in figures.

Ordering and comparing any size numbers

You should know what each of the digits means in a number.
13 720.4

If you are unsure why in this number there are no units or that there are 3 thousands, check your understanding from the section above.

How do you know if a number is bigger or smaller?

12 925.7 and 13 720.4

Here the tens of thousands are the same, 10 000, but the thousands digits are different and the 3 representing 3 thousand is more than the 2 representing 2 thousand, so the second number is larger.

What if it is not that straightforward, for example comparing 9 925.7 and 20 902?

Look carefully at these two numbers. Notice that the first number does not show any ten thousands but the second number shows 2 ten thousands, therefore this number must be the larger. Now the rest of the number isn't used to compare and order the two numbers.

> Begin by looking at the digits starting with the largest part of the number. Find the first ones that are different.

> Note that both numbers have the same number of digits but the first number starts with nine thousand and the second number starts with twenty thousand.

Activity

Try these.

Which number is the larger of each pair?

a 13.7 and 19.2

b 0.75 and 3.2

c 22 and 1.36

d 242 and 2.78

e 93.4 and 24.53

f 1076 and 234.44

Tips for assessment

Know and quickly recognise what different parts of numbers mean, that is, their place value. If there are decimals parts, the decimal point is to the right of the units.

Zeros in a number tell us there is nothing in that place but they are very important to hold that place. So a number 3.08 has 3 units and 8 hundredths, because you have no tenths. So if you are comparing this with 3.8, which has 3 units and 8 tenths, remember that tenths are more than hundredths, so 3.8 is larger than 3.08.

Tasks

1 Write these numbers in words.

  a 236 000

  b 10 042

  c 394.8

  d 0.362

2 Write these numbers in figures.

  a Two hundred and forty

  b Ten point six

  c Twenty-two thousand and forty-five point five

  d One thousand three hundred

3 Write 1.5 million in figures.

4 Show the hundredths part of each of these numbers, then put these numbers in order, largest number first.

  5.679, 12.504, 3.67, 12.461, 5.09

5 **a** Write the following money amounts as dollars using decimals.

  i 13 dollars and 67 cents

  ii 6 dollars 73 cents

  iii 124 dollars 56 cents

  iv 6 dollars and 7 cents (be careful here, look for 2 numbers after the decimal point)

  b Now put the amounts in order, smallest first.

6 Write the following measurements as decimals in centimetres. Remember there are 10 millimetres in a centimetre.

  a 4 centimetres and 6 millimetres

  b 10 centimetres and 3 millimetres

  c 6 centimetres and 8 millimetres

  d 12 centimetres and 6 millimetres

7 Write these measurements as metres. (You may need to look at Unit 102 to remind you about this.)

  a 53 cm and 8 mm

  b 6 m, 60 cm

  c 862 mm

  d 623 cm ← Be careful.

8 Put these decimals in order of size, smallest first.

  a 0.123, 0.456, 0.231, 0.201, 0.102, 0.114

  b 0.871, 0.561, 0.271, 0.914, 0.832, 0.9

  c 0.01, 0.003, 0.1, 0.056, 0.066, 0.008

Test your knowledge

In your exam, the questions are multiple choice. This means that you need to choose the correct answer from the four options given: a, b, c and d. The correct answer will always be shown, but the other answers often look reasonable and may be the result of a simple mistake or an error in the method.

For example, which of these numbers has 6 hundreds?

> **a** 286
>
> **b** 682
>
> **c** 268
>
> **d** 862

All the options have a 6 in them, but only one has 6 in the hundreds column so (b) is the correct answer.

Try these

1 Put these numbers in the order smallest to largest:

34 005   10 002   789   63 254

> **a** 34 005, 10 002, 789, 63 254
>
> **b** 10 002, 789, 63 254, 34 005
>
> **c** 789, 63 254, 34 005, 10 002
>
> **d** 789, 10 002, 34 005, 63 254

2 Put these numbers in the order smallest to largest:

0.564   2.06   0.45   1.232

> **a** 0.564, 2.06, 0.45, 1.232
>
> **b** 0.45, 0.564, 2.06, 1.232
>
> **c** 1.232, 2.06, 0.45, 0.564
>
> **d** 0.45, 0.564, 1.232, 2.06

3 Which of these numbers shows three-thousandths of a metre?

> **a** 0.003 m
>
> **b** 0.03 m
>
> **c** 0.3 m
>
> **d** 3000 m

4 What is one million two hundred and five in figures?

> **a** 1 250
>
> **b** 10 205
>
> **c** 100 025
>
> **d** 1 000 205

5 What is 340 021 in words?

a Thirty-four thousand and twenty-one

b Three million forty thousand and twenty-one

c Three hundred and forty thousand and twenty-one

d Thirty-four hundred and twenty-one

6 What is 50 cents written in dollars?

a $50

b $5.0

c $0.05

d $0.50

7 Put these measurements in order, starting with the smallest:

3.7 m   0.75 m   2.08 m   4.1 m

a 4.1 m, 0.75 m, 3.7 m, 2.08 m

b 0.75 m, 2.08 m, 3.7 m, 4.1 m

c 2.08 m, 3.7 m, 4.1 m, 0.75 m

d 3.7 m, 4.1 m, 0.75 m, 2.08 m

8 In which of these numbers does 5 have the value of five tenths?

a 3.549

b 96.35

c 865.07

d 954.23

9 A report says one million, one hundred and ten thousand Jamaicans have a cellphone subscription.
What is this number in figures?

a 111 000

b 1 110 000

c 11 100 000

d 1 000 110 000

10 What is 700 904 in words?

a Seven hundred nine hundred and four

b Seven thousand nine hundred and four

c Seventy thousand nine hundred and four

d Seven hundred thousand nine hundred and four

Unit 202
Measurement and standard units

Introduction

This unit is about using measure and time to solve everyday problems and builds on the section on measurement in Stage 1 Unit 102.

- Length: measuring how long things are, or how far apart things are
- Weight/mass: measuring how heavy things are
- Capacity: measuring amounts of liquid
- Time: finding what time of day it is, or how long something takes
- Temperature: measuring how hot things are.

Learning objectives

In this unit, you will find information on:

- converting between units in the metric system
- using imperial units
- converting imperial units to metric units
- using units of time in everyday contexts
- using units of temperature in everyday contexts

This will help you prepare for questions about:

- comparing amounts in different units or systems
- solving problems related to length, weight, capacity or temperature
- planning and scheduling activities
- working out time differences between countries.

Measuring and estimating length

In Stage 1 we looked at using scales for measurement of length, weight and capacity using metric measurements. You may wish to reread this now before moving on to converting between metric measurements.

Length is a measure of how long things are.

We also use these words

- **distance** – for how far apart things are
- **width** – for how wide things are
- **height** – for how tall or high things are.

We need to measure length accurately in many situations:

- knowing that items will fit in a space, for example fitting a bed into a room
- cutting materials to make things, for example a piece of wood to make a shelf
- knowing how far to travel to get to a place.

Common units for the distance between places are miles and kilometres.

A mile is roughly the distance an average person could walk in 20 to 25 minutes.

An average person could walk one kilometre in 10 to 15 minutes.

A car could travel one mile in a minute on a fast road.

Common units for length are millimetres, centimetres and metres.

A millimetre is a very small length. It is roughly the thickness of a coin.

A millimetre is often written as mm, so 30 mm is short for 30 millimetres.

We often use millimetres to measure very short lengths, such as a 12 mm screw.

Builders and engineers also measure longer lengths in millimetres to be very precise.

There are 10 millimetres in 1 centimetre.

There are 1000 millimetres in 1 metre.

A centimetre is roughly the width of a little finger.

A centimetre is often written as cm, so 25 cm is short for 25 centimetres.

We use centimetres to measure everyday items like furniture.

There are 100 centimetres in 1 metre.

A centimetre is the same as 10 millimetres.

A metre is roughly the length of one stride.

A metre is often written as m, so 5 m is short for 5 metres.

We use metres to measure longer lengths such as a room or a running race.

One metre is the same as 100 centimetres.

One metre is also the same as 1000 millimetres.

Converting lengths

The same measurement may be expressed as 1.8 metres, 1 m 80 cm, 180 cm or 1800 mm, so we need to be able to convert accurately between lengths given in metres, centimetres and millimetres.

To do this, you need to be able to multiply and divide by 10, 100 and 1000.

See the topics on multiplication and division for more on this in Units 105 and 106.

To convert millimetres to centimetres, divide by 10

500 ÷ 10 = 50, so 500 millimetres is the same as 50 centimetres.

35 ÷ 10 = 3.5, so 35 millimetres is the same as 3.5 centimetres.

To convert millimetres to metres, divide by 1000

2400 ÷ 1000 = 2.4, so 2400 millimetres is the same as 2.4 metres.

800 ÷ 1000 = 0.8, so 800 millimetres is the same as 0.8 metres.

To convert centimetres to millimetres, multiply by 10

$92 \times 10 = 920$, so 92 centimetres is the same as 920 millimetres.

$6.4 \times 10 = 64$, so 6.4 centimetres is the same as 64 millimetres.

To convert centimetres to metres, divide by 100

$4700 \div 100 = 47$, so 4700 centimetres is the same as 47 metres.

$60 \div 100 = 0.6$, so 60 centimetres is the same as 0.6 metres.

To convert metres to millimetres, multiply by 1000

$1.2 \times 1000 = 1200$, so 1.2 metres is the same as 1200 millimetres.

$8 \times 1000 = 8000$, so 8 metres is the same as 8000 millimetres.

To convert metres to centimetres, multiply by 100

$7.3 \times 100 = 730$, so 7.3 metres is the same as 730 centimetres.

$12 \times 100 = 1200$, so 12 metres is the same as 1200 centimetres.

Try it out: convert 107 millimetres to metres.

This is converting millimetres (small units) to metres (larger units), so divide.

The conversion rate is 1000 mm = 1 m.

Divide by 1000.

$107 \div 1000 = 0.107$

So 107 millimetres is the same as 0.107 metres.

> **Learner tip**
>
> It's useful to think of a rule for remembering conversions easily.
>
> Converting from a small unit to a larger unit, divide.
>
> Converting from a large unit to a smaller unit, multiply.
>
> Learn the conversions:
>
> 10 mm = 1 cm
>
> 100 cm = 1 m
>
> 1000 mm = 1 m

Activity

Complete the table to show lengths in millimetres, centimetres and metres.

The first one is done for you.

2500 mm	250 cm	2.5 m
	25 cm	0.25 m
80 mm		0.08 m
	2500 cm	25 m
8000 mm	800 cm	
110 mm		0.11 m
		1.1 m
11000 mm		
	2.5 cm	0.025 m

Measuring using scales

At Stage 1, we looked at measuring lengths using a ruler or tape measure.

This diagram shows a ruler marked in centimetres.

> The numbered divisions are centimetres and the smaller unlabelled ones are millimetres.

The pointer is between 7 cm and 8 cm.

It is pointing at the 4th small division after 7 cm, which is 7 cm and 4 mm.

We can write this as 7.4 cm or 74 mm.

This diagram shows part of a measuring tape marked in centimetres.

> The numbered divisions are centimetres, the small unlabelled ones are millimetres.

The pointer is between 143 cm and 144 cm.

It is pointing at the 9th small division after 143 cm, which is 143 cm and 9 mm.

This is written as 143.9 cm or 1 439 mm.

Example

Kai measures her sitting room.

The pointer on this scale shows the length of one wall.

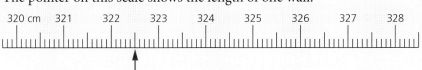

She wants to fit a sofa and a cupboard along this wall.

The sofa is 1.8 metres long.

The cupboard is 1.3 metres long.

What is the length of the wall?

Is there enough space along the wall for the sofa and the cupboard?

Solution

The pointer shows 322.5 centimetres.

$1.8 + 1.3 = 3.1$

The total length of the sofa and cupboard is 3.1 metres.

$3.1 \times 100 = 310$

The wall is 322.5 cm, the sofa and cupboard are 310 cm, so there is enough space.

> **Tips for assessment**
> Always give the correct units with an answer.

Measuring and estimating weight

Weight is a measure of how heavy things are.

We need to measure weights accurately in many situations:

- working out amounts of ingredients for cooking
- paying the right amount for sending parcels or taking bags on planes
- checking body weight to see if a person is healthy.

Common units for weight are grams and kilograms.

A gram is a very small weight. A peanut weighs one gram.

A gram is often written as g, so 50 g is short for 50 grams.

There are 1000 grams in 1 kilogram.

A kilogram is a heavier weight. A bag of sugar weighs one kilogram.

A kilogram is often written as kg, so 25 kg is short for 25 kilograms.

One kilogram is the same as 1000 grams.

We use grams for lighter weights and kilograms for heavier weights.

Converting weights

The same weight may be expressed as 1.5 kg, 1 kg and 500 g or 1 500 g so we need to be able to convert accurately between weights given in grams and weights given in kilograms.

There are 1000 grams in a kilogram.

To convert grams to kilograms, divide the number of grams by 1000

$3000 \div 1000 = 3$, so 3000 grams is the same as 3 kilograms.

$2500 \div 1000 = 2.5$, so 2500 grams is the same as 2.5 kilograms.

$800 \div 1000 = 0.8$, so 800 grams is the same as 0.8 kilograms.

To convert kilograms to grams, multiply the number of kilograms by 1000

$2 \times 1000 = 2000$, so 2 kilograms is the same as 2000 grams.

$1.2 \times 1000 = 1200$, so 1.2 kilograms is the same as 1200 grams.

$0.3 \times 1000 = 300$, so 0.3 kilograms is the same as 300 grams.

Activity

Copy and complete the table to show weights in grams and in kilograms.
The first one is done for you.

4000 g	4 kg
	0.4 kg
320 g	
3200 g	
	0.032 kg
40 g	
1500 g	
	0.15 kg
	0.015 kg
	0.001 kg

Using weighing scales

At Stage 1, we looked at measuring weight using scales on different weighing instruments.

This diagram shows the sort of scale found on weighing machines.

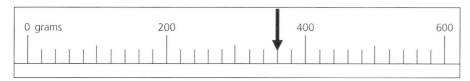

The pointer shows the weight of the item.

The pointer is between 200 grams and 400 grams.

To work out the exact weight, we need to know what weight the divisions show (shown by the small lines).

There are 10 divisions between 200 grams and 400 grams.

Those 10 divisions cover 200 grams.

So, each division is 20 grams. ←

You can check by counting 20s for each division. The first division after 200 grams is 220 grams, then the next is 240 grams and so on up to 400.

There are 8 divisions between 200 grams and the pointer.

8 divisions are 160 grams.

160 grams more than 200 is 360 grams – that is the weight shown.

To check, there are 2 divisions between the pointer and 400 grams.

2 divisions are 40 grams.

40 grams less than 400 grams is 360 grams.

> On different weighing machines, the unlabelled divisions might not be 10 grams.
> They could be 5 grams, 20 grams, 50 grams, 250 grams 0.1 kg, 2 kg etc.

Activity

Work out what each small division is on these scales.

a 0 g   250

b 0 kg   50

c 0 g   1000

d 0 kg   10

Example

Amanda wants to lose weight.

The pointer on this scale shows her weight.

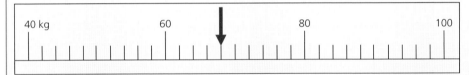

Her target weight is 65 kg.

How much more weight should she lose to meet her target weight?

Solution

The pointer shows a weight between 60 kg and 80 kg.

There are 10 divisions between 60 kg and 80 kg, so each division is 2 kg.

There are 4 divisions from 60 kg to the pointer.

4 divisions are 8 kg.

The weight is 60 + 8 = 68 kg.

Her target weight is 65 kg.

68 − 65 = 3

So she should lose 3 kg.

Measuring and estimating capacity

Capacity is a measure of the amount a container can hold.

We need to measure capacity accurately in many situations:

- measuring liquids for cooking
- knowing how much liquid there is in a container, for example a can of oil
- knowing how much space there is in a fridge or a car boot.

We measure capacity in litres.

A litre is the amount of fruit juice in a standard carton.

A litre can be written as ℓ, so $2\,\ell$ is short for 2 litres.

For smaller amounts, we use millilitres.

A millilitre is a very small amount, roughly the same as a few drops of water.

A millilitre is often written as ml, so 100 ml is short for 100 millilitres.

There are 1000 millilitres in 1 litre.

We use millilitres for smaller amounts and litres for larger amounts.

Converting capacities

We need to be able to convert accurately between capacities given in millilitres and litres.

There are 1000 millilitres in a litre.

To convert millilitres to litres, divide the number of millilitres by 1000.

$5000 \div 1000 = 5$, so 5000 millilitres is the same as 5 litres.

$2400 \div 1000 = 2.4$, so 2400 millilitres is the same as 2.4 litres.

$500 \div 1000 = 0.5$, so 500 millilitres is the same as 0.5 litres.

To convert litres to millilitres, multiply the number of litres by 1000.

$3 \times 1000 = 3000$, so 3 litres is the same as 3000 millilitres.

$2.2 \times 1000 = 2200$, so 2.2 litres is the same as 2200 millilitres.

$0.75 \times 1000 = 750$, so 0.75 litres is the same as 750 millilitres.

Activity

Copy and complete the table to convert the capacities.

The first one is done for you

7000 ml	7 ℓ
700 ml	
	0.95 ℓ
	9.5 ℓ
95 ml	
	0.07 ℓ

Using scales

At Stage 1, we looked at measuring capacity using scales on different weighing instruments.

This diagram shows the sort of scale found on measuring jugs.

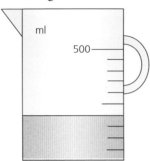

The level of the liquid is between 0 ml and 500 ml.

There are 10 divisions between 0 ml and 500 ml.

Those 10 divisions cover 500 ml.

So, each division is 50 ml.

The level is 4 divisions above 0 ml.

$4 \times 50 = 200$

4 divisions are 200 ml – that is the amount of liquid in the jug.

This diagram shows a measuring jug with a different scale.

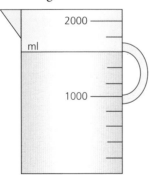

In different measuring containers, the divisions might not be 200 ml. They could be 20 ml, 50 ml, 100 ml, 500 ml, 2 litres, 5 litres, etc.

The level of the liquid is between 1000 ml and 2000 ml.

There are 5 divisions between 1000 ml and 2000 ml.

Those 5 divisions cover 1000 ml.

So, each division is 200 ml.

The level of the liquid is 3 divisions above 1000 ml.

$3 \times 200 = 600$

The amount of liquid is 1600 ml.

Activity

Work out what each division is on these scales.

a

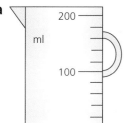

b

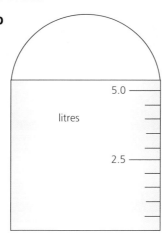

c

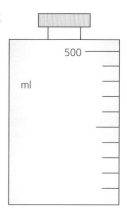

Example

A mechanic needs 300 ml of oil to put in an engine.

He pours some oil from this measuring bottle.

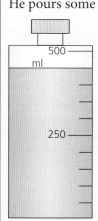

How much oil will be left after he has put 300 ml in the engine?

Solution

The level is between 250 ml and 500 ml.

There are 5 divisions between 250 ml and 500 ml.

Those 5 divisions cover 250 ml.

Each division is 50 ml.

The level is 4 divisions above 250 ml.

$4 \times 50 = 200$

$250 + 200 = 450$

There are 450 ml of oil in the bottle.

The mechanic pours out 300 ml.

$450 - 300 = 150$

So there will be 150 ml left in the bottle.

Tasks

1 What is the length shown by the pointer?

2 What is the length shown by the pointer?

3 Levi cuts a piece of wood to make some shelves.

The shelves will be 800 millimetres long.

The pointer on this scale shows the length of the piece of wood.

    a What is the length of the piece of wood?

    b How many shelves can he cut from the piece of wood?

4 Margaret wants to put up fencing on one side of a garden.

The pointer on this scale shows the length of the side of the garden.

One fence panel is 1.2 metres long including the posts.

    a What is the length of the side of the garden?

    b How many fence panels will Margaret need?

5 What is 3 650 mm expressed as

    a metres

    b centimetres?

6 What is 1 400 grams in kilograms?

7 What is 0.16 kilograms in grams?

8 What is the reading on this scale?

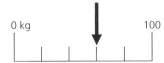

9 A mail order company sells candles.
One large candle weighs 350 g.
A customer orders 8 large candles.
What is the total weight of the candles?

10 Amber is going on holiday.

She weighs her luggage.

The pointer on this scale shows the weight.

Amber can take a maximum of 23 kg of luggage.

Amber must pay $10 for every kilogram over this weight.

How much must she pay for her luggage?

11 What is 37 000 millilitres in litres?

12 What is 1.8 litres in millilitres?

13 How much liquid is there in this bottle?

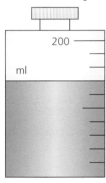

14 How much liquid is there in this container?

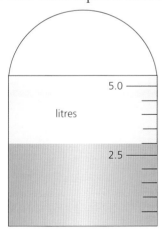

15 A gardener uses a liquid grass food on a lawn.

He pours 600 ml of liquid from this container.
How much liquid will be left in the container
after he has poured out 600 ml?

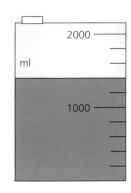

Using imperial units

At Stage 2 you will be expected to understand measurements in imperial units and use given conversions.

Length

The metric units of length are millimetres (mm), metres (m) and kilometres (km).

1000 millimetres = 1 metre

1000 metres = 1 kilometre

Common imperial units of length are inches (in), feet (ft), yards (yd) and miles (mi).

12 inches = 1 foot

3 feet = 1 yard

> ### Activity
>
> A foot was supposedly the length of a person's foot and an inch is approximately the length of the first joint on your index finger. One yard is roughly the distance from your nose to the first knuckle of your fingers on an outstretched arm.
>
> See how close these approximations are to your own measurements.

Inches are smaller than feet, so if you need to find the number of inches this will be a multiplication calculation.

Yards are bigger than feet, so if you need to find the number of yards this will be a division calculation.

> ### Example
> **a** How many inches are there in 4 feet? (12 inches = 1 foot) At Stage 2, you will be given the conversion rate in the question.
> **b** A roll of material has 30 ft of material.
> How many yards is this? (3 feet = 1 yard).
>
> ### Solution
> **a** 12 inches = 1 foot is the same as saying in 1 foot there are 12 inches, so you need 4 × 12 for 4 ft.
> 4 × 12 = 48
> There are 48 inches in 4 feet.
> **b** There are 3 ft in 1 yd and the roll has 30 ft, so you need to work out 30 ÷ 3 for yards.
> 30 ÷ 3 = 10
> There are 10 yards of material on the roll.

Weight

The metric units of weight are grams (g), kilograms (kg) and tonnes (t).

1000 grams = 1 kilogram

1000 kilograms = 1 tonne

Common imperial units of weight are ounces (oz), pounds weight (lb) and stones (st).

16 ounces = 1 pound weight

14 pounds weight = 1 stone

Example

a How many ounces are in 2 lb? (16 ounces = 1 pound weight)

b A child weighs 36 pounds. How many stone is this?
(14 pounds = 1 stone) You will be given the relevant conversion rate.

Solution

a 16 ounces = 1 pound is the same as saying in 1 lb there are 16 oz, so you need 2 × 16 for 2 lb

2 × 16 = 32. There are 32 ounces in 2 pounds.

b There are 14 lb in 1 st and the child weighs 36 lb, so you need

36 ÷ 14 for stones

36 ÷ 14 = 2 st and 8 left over. The child weighs 2 st and 8 lb.

> Ounces are smaller than pounds, so if you need to find the number of ounces this will be a multiplication calculation.

> A stone is heavier than a pound, so if you need to find the number of stones this will be a division calculation.

Capacity

The metric units of capacity are millilitres (ml), litres (ℓ) and cubic metres (m^3).

1000 millilitres = 1 litre

1000 litres = 1 cubic metre

Common imperial units of capacity are fluid ounces (fl. oz.), pints (pt) and gallons (gal). You will only work with pints and gallons at Stage 2.

8 pints = 1 gallon

Example

a How many pints are there in 4 gallons? (8 pints = 1 gallon)

b Convert 40 pints into gallons. (8 pints = 1 gallon)

Solution

a 8 pints = 1 gallon is the same as saying in 1 gal there are 8 pt, so you need 4 × 8 for 4 gallons

4 × 8 = 32

There are 32 pints in 4 gallons.

b There are 8 pt in 1 gal, so you need 40 ÷ 8 for gallons.

40 ÷ 8 = 5

40 pints equals 5 gallons.

> Pints are smaller than gallons, so if you need to find the number of pints this will be a multiplication calculation.

> Gallons are larger than pints, so if you need to find the number of gallons this will be a division calculation.

Converting imperial units to metric units

At Stage 2, you will be expected to convert from imperial measurements to metric measurement using approximations that will be given to you.

Length

Common approximate conversion rates:

1 inch is slightly more than 2.5 centimetres.

1 yard is slightly less than 1 metre.

Five-eighths of a mile is approximately equal to 1 kilometre.

Weight

Common approximate conversion rate:

2 pounds is slightly less than 1 kilogram.

Capacity

Common approximate conversion rate:

2 pints is slightly more than 1 litre.

For calculations, don't worry whether the conversion rate is 'slightly more' or 'slightly less', just use the figures you are given.

> Centimetres are smaller than inches, so if you need to find the number of centimetres this will be a multiplication calculation.

> Kilograms are larger than pounds, so if you need to find the number of kilograms this will be a division calculation.

> Litres are larger than pints, so if you need to find the number of litres this will be a division calculation.

Example

a How many centimetres are in 10 inches?

b How many kilograms are in 10 pounds?

c A recipe requires 2 pints of stock. How much is this in litres?

Solution

a If 1 inch = 2.5 cm then you need 10 × 2.5 for 10 inches.

10 × 2.5 = 25.

There are approximately 25 centimetres in 10 inches.

b 2 pounds is slightly less than 1 kilogram, so you need 10 ÷ 2 for kilograms.

10 ÷ 2 = 5

10 pounds is approximately 5 kilograms.

c 2 pints is slightly more than 1 litre, so you need 2 ÷ 2 for litres.

2 ÷ 2 = 1

2 pints is approximately 1 litre.

Tasks

1 How many pints are there in 3 gallons? (8 pints = 1 gallon)

2 How many ounces are in $2\frac{1}{2}$ lb? (16 ounces = 1 pound weight)

3 How many inches are in 2 yards? (12 inches in 1 foot and 3 feet in 1 yard)

4 A piece of material is 5 yards long. The customer needs 5 metres. Is the material long enough?

5 Approximately how many kilometres are there in 5 miles?

6 A cook has 3 lb of flour. Approximately how many kilograms is this?

7 Approximately how many litres are there in 6 pints?

8 A pen is 6 inches long. Approximately how many centimetres is this?

Using units of time

Measuring time is part of everyday life at home, work and leisure.

This includes

- reading clocks and stop watches
- understanding times expressed in different ways
- using a timetable to plan a journey
- converting between units of time, for example hours and minutes
- calculating lengths of time, for example the length of a working day.

At Stage 2, you will build on your knowledge of reading analogue and digital clocks from Stage 1 and become familiar with **second hands** and stopwatches.

Sometimes clocks only have two hands: the shorter hand, or **hour hand**, and the longer hand, or **minute hand**.

This clock face has three hands.

> When the minute hand points at the number 12 this is known as o'clock.

> The shortest hand is the hour hand, which is just past eight o'clock.

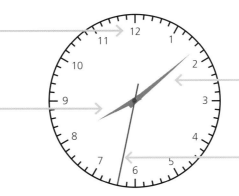

> The second longest hand is the minute hand. If we count the minutes past o'clock, there are 5 small tick marks to the number 1 and then 2 more making 7 minutes. If you are unsure of this have another look at Unit 102 (page 25) The time is seven minutes past eight and is written 8:07.

> The longest hand is called the second hand as it shows the vseconds. Sometimes this is another colour such as red and it is usually thinner than the other hands.
>
> Where is the second hand pointing? This is at 2 small tick marks past 6. We know that 6 is usually half-past or 30 minutes past and as this is the second hand it is showing 30 seconds past. Then we add the two tick marks past 30 and the total is 32 seconds. We could count up all the small tick marks past 12 but this would take a long time.
>
> This means that 32 seconds have passed since this minute started and there are 27 seconds until it is 8:08. The time is shown as 8:07:32.

Is this morning or evening? Do you remember a.m. and p.m. from Stage 1? If this is in the morning we say 8:07:32 a.m. and if it is evening we say 8:07:32 p.m.

A stopwatch can be used for timing a short period of time such as running a race. This one is an analogue stopwatch. If it is used for timing a race, it is started when the race starts by pushing the button and stopped when the race finishes. This makes it easy to time short events.

How many seconds have passed?

It is showing 10 seconds.

This is a digital stopwatch. It shows minutes (min) and seconds (sec).

How many minutes and seconds have passed since the stopwatch was started?

14 minutes and 26 seconds have passed.

Units of time

You are expected to know this information:

60 seconds = 1 minute

60 minutes = 1 hour

24 hours = 1 day

7 days = 1 week

52 weeks = 1 year

365 days = 1 year (or 366 days in a leap year)

12 months = 1 year

The number of days in a month varies between 28 and 31. You may know this rhyme:

30 days has September,

April, June, and November.

When short February's done

All the rest have 31.

Or this one:

Thirty days has September,

April, June, and November,

All the rest have thirty-one.

February has twenty-eight,

But leap year coming one in four

February then has one day more.

Converting between units

The usual rule for converting is:

- converting from a small unit to a larger one, divide
- converting from a large unit to a smaller one, multiply.

With measurements of time, we need to adapt this.

We usually calculate with tens, hundreds and thousands, but with time we calculate with other numbers such as 60. ←

> Converting hours to minutes, a large unit to a smaller one, we could multiply by 60.

$4 \times 60 = 240$, so 4 hours is 240 minutes.

To convert 2 hours 35 minutes to minutes, do not calculate 2.35×60.

2.35 hours is not the same as 2 hours 35 minutes.

This is a common mistake.

It's best to convert the whole hours, then add the minutes.

2 hours is $2 \times 60 = 120$ minutes.

$120 + 35 = 155$

2 hours 35 minutes is 155 minutes.

You should also know common fractions of an hour.

A quarter of an hour = 15 minutes

Half an hour = 30 minutes

Three-quarters of an hour = 45 minutes

Use this information to convert $2\frac{1}{2}$ hours to minutes.

2 hours $\times 60 = 120$ minutes

Half an hour is 30 minutes.

$120 + 30 = 150$ minutes

Converting minutes to hours, a small unit to a larger one, we could divide by 60.

There is no problem where the number of minutes divides exactly by 60.

$120 \div 60 = 2$, so 120 minutes is 2 hours.

Where the number of minutes does not divide by 60 exactly, it's more complicated.

$140 \div 60 = 2.33$ recurring, so 140 minutes is roughly 2.33 hours.

This is not very useful. It is easy to confuse 2.33 hours with 2 hours 33 minutes, which is not the same.

It is better to use the subtraction method to divide with time.

$140 - 60 = 80$

$80 - 60 = 20$

140 minutes is 2 hours (we've taken 60 away twice). There are 20 minutes left.

140 minutes is 2 hours 20 minutes.

▲ Montego Bay

Using a timetable

This is an extract from a timetable for some buses from Montego Bay to Falmouth.

Montego Bay	Falmouth
5.10 a.m.	5.40 a.m.
7.10 a.m.	7.40 a.m.
9.10 a.m.	9.40 a.m.
11.10 a.m.	11.40 a.m.
1.10 p.m.	1.40 p.m.
2.10 p.m.	2.40 p.m.

1 What time does the 9.10 a.m. bus from Montego Bay arrive at Falmouth?

2 The time is 6.50 a.m. What time is the next bus to Falmouth?

3 What time does the bus that arrives in Falmouth at twenty minutes to two leave Montego Bay?

4 Which is the latest bus from Montego Bay to get to Falmouth before 1 p.m.?

Adding and subtracting time

Adding and subtracting time is different to working with other numbers.

This is because with time we are not dealing with tens, hundreds and thousands.

Time is based on other numbers, as there are 60 minutes in an hour.

Example 1

Here's a time problem and a way of solving it.

A football match starts at 2.45 p.m.

It will last 1 hour 40 minutes including a half-time break.

At what time should it end?

Solution

To answer this problem, don't just add 2.45 and 1.40.

This will give an incorrect answer – but it is a common error.

It's best to use a counting up method, adding the hours and minutes separately.

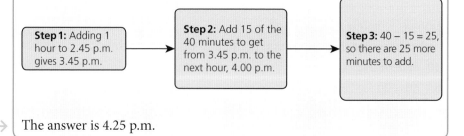

Step 1: Adding 1 hour to 2.45 p.m. gives 3.45 p.m.

Step 2: Add 15 of the 40 minutes to get from 3.45 p.m. to the next hour, 4.00 p.m.

Step 3: 40 – 15 = 25, so there are 25 more minutes to add.

It is important to add p.m.

The answer is 4.25 p.m.

Example 2

This is a time problem involving finding a difference.

The method is similar.

Thomas starts work at 8.15 a.m.

His journey from home to work takes 1 hour 20 minutes.

At what time should he leave home?

Solution

To answer this problem, don't just subtract 1.20 from 8.15.

This will give an incorrect answer – but it is a common error.

It's best to deal with the hours and minutes separately.

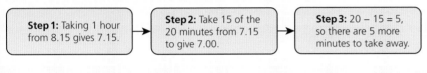

Step 1: Taking 1 hour from 8.15 gives 7.15. → **Step 2:** Take 15 of the 20 minutes from 7.15 to give 7.00. → **Step 3:** 20 – 15 = 5, so there are 5 more minutes to take away.

The answer is 6.55 a.m. ← It is important to add a.m.

Example 3

Thomas is at work from 8.15 a.m. to 4.45 p.m.

For how many hours is he at work?

Solution

To answer this problem, don't just subtract 8.15 from 4.45 on a calculator.

This will give an incorrect answer – but it is a common error.

Count up from 8.15 in steps.

Step 1: 8.15 a.m. to 9.00 a.m. is 45 minutes.

Step 2: 9.00 a.m. to 12.00 p.m. is 3 hours.

Step 3: 12.00 p.m. to 4.00 p.m. is 4 hours.

Step 4: 4.00 p.m. to 4.45 p.m. is 45 minutes.

Add together the times from all the steps:

hours first: 3 + 4 = 7 hours

minutes: 45 + 45 = 90 minutes

90 – 60 = 30, so 90 minutes is 1 hour 30 minutes.

7 hours + 1 hour 30 minutes is 8 hours 30 minutes.

The answer is 8 hours 30 minutes, or $8\frac{1}{2}$ hours. ← It is important to include the units.

Tasks

1 Try some conversions between times in minutes and the same times in hours and minutes.

The first one is done for you.

Hours and minutes	The same time in minutes
1 hour 20 minutes	80 minutes
	325 minutes
3 hours 10 minutes	
2 hours 35 minutes	
6 hours 5 minutes	
	270 minutes
3 hours 50 minutes	
	135 minutes
	195 minutes
$2\frac{1}{2}$ hours	
	105 minutes

2 What time is $2\frac{1}{2}$ hours before 6.05 a.m.?

3 A group of friends want to see a film at the cinema.
The film starts at 20:10.
The film lasts 145 minutes.
What time will the film end?

4 A plumber repairs a boiler.

He starts the job at 8.45 a.m. and finishes it at 2.15 p.m.

Fill in the details on this time sheet.

	Time (24-hour clock)
Start of job	08:45
End of job	

Total time taken (hours and minutes)	

5 Margaret is going home to Jamaica.

She wants to take a bus to Heathrow Airport. She uses this timetable.

Torquay	...	...	0520	0700	...	0940	1140	1400
Newton Abbot	...	...	0535	...	...	1000	1200	1420
Exeter	0140	0435	0615	0830	0945	1045	1245	1500
Heathrow Airport	0530	0805	0950	1150	1305	1410	1610	1850
London Victoria	0615	0920	1050	1250	1405	1510	1710	1950

She wants to go from Exeter to Heathrow Airport.
She wants to arrive by 1.30 p.m.

a What is the latest time she can catch a bus from Exeter?

b How long will the bus journey take?

Using units of temperature

Temperature is a measure of how hot things are.

We need to measure temperature accurately in many situations:

- finding body temperature to see if a person is well
- finding the right temperature for storing and cooking food safely
- finding air temperature so that we wear suitable clothes.

Temperature is measured in degrees **Celsius**, often written as °C, or in degrees Fahrenheit, often written as °F. At Stage 1 we used Fahrenheit. At Stage 2 you will use degrees Celsius.

90 °F means 90 degrees Fahrenheit.

32 °C means 32 degrees Celsius.

You will not be required to know how to convert between degrees Celsius and Fahrenheit.

Here are some common temperatures.

	Fahrenheit	Celsius
Water freezes at	32 °F	0 °C
A typical room temperature is	72 °F	22 °C
Normal human body temperature is	98–99 °F	37 °C
Water boils at	212 °F	100 °C

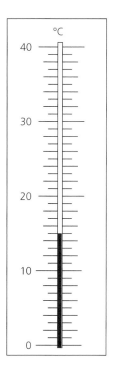

Measuring temperature

We use a thermometer to measure temperature.

There are many types of thermometer. For more on the basics of reading a temperature from a thermometer, see Unit 102.

At Stage 2, you may be expected to read thermometers with different scales.

Look at the thermometer on the right.

There are 10 divisions between 10 degrees and 20 degrees.

Those 10 divisions cover 10 degrees.

So each division is 1 degree.

The top of the column is 5 divisions above 10 degrees.

The temperature is 15 degrees Celsius or 15 °C.

Example

This diagram shows a different thermometer. What is the temperature?

Solution

The red bar is between 37.0 °C and 38.0 °C.

There are 10 divisions between 37.0 °C and 38.0 °C.

Those 10 divisions cover 1.0 degrees.

Each division is one tenth of a degree, or 0.1 °C.

The red bar is at the 7th division after 37 °C.

$7 \times 0.1 = 0.7$

$37.0 + 0.7 = 37.7$

The temperature on the display is 37.7 °C.

Tasks

1 This thermometer shows a child's body temperature.

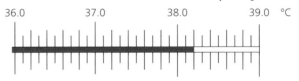

The child has a fever if the temperature is 37.5 °C or higher.

a What is the temperature shown on the thermometer?

b What value does each small division have?

c Does the child have a fever?

2 What is the temperature on this cooking thermometer?

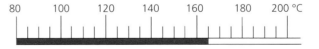

Tasks

3 What is the temperature on this thermometer?

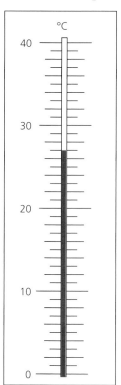

4 What is the temperature on this thermometer?

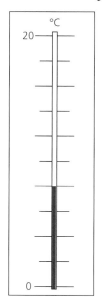

5 A nurse takes a patient's temperature.

This thermometer shows the temperature.

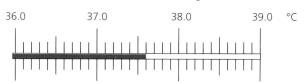

The previous day, the temperature was 37.9 °C.

What is the change in temperature since the previous day?

Test your knowledge

1 What is the weight shown on this scale?

   a 596 grams

   b 560 grams

   c 486 grams

   d 480 grams

2 What is 1200 mm in metres?

   a 0.12 m

   b 1.2 m

   c 12 m

   d 120 m

3 What is the temperature on this cooking thermometer?

   a 143 °C

   b 146 °C

   c 155 °C

   d 158 °C

4 What is 225 minutes in hours and minutes?

   a 3 hours 45 minutes

   b 3 hours 35 minutes

   c 2 hours 15 minutes

   d 2 hours 25 minutes

5 A man sets off on a car journey at 9:30 a.m.
The journey takes $2\frac{3}{4}$ hours.
What time should he arrive?

   a 11.15 a.m.

   b 12.15 p.m.

   c 12.05 p.m.

   d 12.25 p.m.

6 What is 4 feet in inches? (12 inches equals 1 foot)

   a 3 inches

   b 12 inches

   c 30 inches

   d 48 inches

7 What is the amount of liquid in this measuring jug?

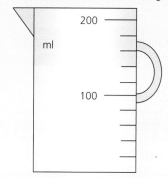

   a 104 ml

   b 110 ml

   c 120 ml

   d 140 ml

8 Jo wants to take a 13:20 flight.

   She wants to get to the airport 90 minutes before the flight.

   What time should she get to the airport?

   a 11.50 a.m.

   b 12.50 p.m.

   c 1.50 p.m.

   d 2.50 p.m.

9 How many 500 g bags can a pet shop owner fill from a 15 kg sack?

   a 3

   b 30

   c 33

   d 7500

10 Approximately how many centimetres are there in 10 inches?

   a 4 cm

   b 5 cm

   c 10 cm

   d 25 cm

Introduction

This unit builds on the skills in adding, subtracting, multiplying and dividing whole numbers you developed at Stage 1 in Unit 105.

There is no substitute for practice when it comes to improving speed and confidence in adding and subtracting quickly and efficiently. Your multiplication tables from Stage 1 will come in useful for multiplication and division. You should know the 2, 3, 4, 5, 6, 7, 8, 9 and 10 multiplication tables well enough to be able to recall them quickly.

Learning objectives

In this unit you will find information on:

- calculations with whole numbers up to one million
- multiplication of a number with up to four digits by a number with one or two digits
- division of a number with four digits by a number with one or two digits where the answer may not be a whole number.

This will help you to prepare for questions about:

- calculations with a combination of two operations from addition, subtraction, multiplication and division to solve problems without a calculator.

Addition and subtraction

At Stage 1 you looked at pairs of numbers that add up to 10. It is important to learn pairs of numbers which make 10, 20 and 100 and be familiar with decimal numbers in money.

You will be much happier learning the rest of the mathematical ideas at this level if you are confident and quick at adding and subtracting whole and decimal numbers, so practise whenever you get the chance.

Activity

In pairs, ask each other addition and subtraction questions with pairs of numbers

a which make 10

b which make 20

c which make 100.

Written methods

You should already be confident in using written methods for adding and subtracting whole numbers, but it is always worth recapping on this important skill.

Example 1

Use column addition to work out 38 + 126.

Solution

$$
\begin{array}{r}
3\ 8 \\
+\ 1\ 2\ 6 \\
\hline
1\ 6\ 4 \\
\scriptstyle 1
\end{array}
$$

First set the addition in columns.

Add the digits in the units column 8 + 6 = 14, write 4 in the units column and a small 1 under the tens column (to show you have an extra 10 to add).

Add the digits in the tens column 3 + 2 = 5, 5 + 1 = 6

There is only 1 digit in the hundreds column, so write this 1.

38 + 126 = 164

> Make sure you line up the units, tens and hundreds correctly. These examples show column addition and subtraction. You may use a slightly different layout or method and if it works reliably that is fine. Just check that you get the same answer.

Example 2

Use column subtraction to work out 164 − 38.

Solution

$$
\begin{array}{r}
1\ {}^{5}\!\!\!\diagup\!\!6\ {}^{1}4 \\
-\ \ \ 3\ 8 \\
\hline
1\ 2\ 6
\end{array}
$$

First set the subtraction in columns.

Subtract the digits in the units column.

You cannot take 8 units from 4 units so you will need to use partitioning and split the number differently. Instead of 100 + 60 + 4, 164 becomes 100 + 50 + 14. Change the 6 tens to 5 tens to show that you have moved 1 ten to the units column and change the 4 units to 14 units. Now you can subtract the units (14 − 8 = 6). Write 6 in the units column.

Subtract the digits in the tens column (5 tens − 3 tens = 2 tens).

Write 2 in the tens column.

There is only 1 digit in the hundreds column, so write this (1).

164 − 38 = 126

The best one to use will change with the numbers or the context, but you should aim to be able to use at least two different methods quickly and confidently so you can choose the best one to use.

Real world maths

Practise mental calculations whenever you can such as when you are shopping. Try to add the cost of each item you pick up so that you have an overall idea of the final bill when you get to the cashier. The more you practise, the more likely your total will match the cashier's calculations.

Tips for assessment

Instant recall of number pairs to 10, 20, and 100 will help you use this method quickly and accurately.

Learner tip

An alternative method to answer this question is to add the complement to 100 of 59, which is 41, to 32 to get 73. The most efficient method will vary depending on the numbers.

Mental methods

In this section, you will work through a range of mental calculation methods.

Partitioning with whole numbers

In partitioning, a number is split into its separate parts, for example 456 is split into $400 + 50 + 6$. This helps us to add and subtract numbers, for example:

a Addition $123 + 456$
Split the second number into hundreds, tens and units
$(456 = 400 + 50 + 6)$.
Add the parts in turn $(123 + 400 = 523; 523 + 50 = 573; 573 + 6 = 579)$.
$123 + 456 = 579$

b Subtraction $545 - 324$
Split the second number into hundreds, tens and units $(324 = 300 + 20 + 4)$
Subtract the parts in turn $(545 - 300 = 245; 245 - 20 = 225; 225 - 4 = 221)$
$545 - 324 = 221$

Complements with whole numbers

This is when we use an easier number close to the one we are adding or taking away and then adjust the answer.

For example, to work out $238 + 94$:

Find the complement to 100 of 94 (6).

Complete the equivalent 2-part calculation: $238 + 100 = 338; 338 - 6 = 332$.

So $238 + 94 = 332$

To work out $312 - 67$:

Find the complement to 100 of 67 (33).

Complete the equivalent 2-part calculation: $312 - 100 = 212; 212 + 33 = 245$

So $312 - 67 = 245$

Counting on with whole numbers

Counting on is useful when subtracting.

To work out $132 - 59$, use a physical or mental number line to find the route between 59 and 132.

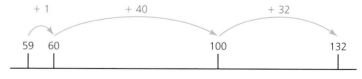

You count on $+1$, $+40$ and $+32$.

So add $1 + 40 + 32 = 73$

Rounding

Subtract 9 from 33.

Round 9 to 10 and take 10 from 33. $33 - 10 = 23$

You have taken away 1 too many, so compensate by adding 1 to 23.

$23 + 1 = 24$ and the answer is 24.

> **Activity**
>
> Use your ruler to draw a line which is 5.6 cm long.
>
> Mark on the line a point which is 2.8 cm from one end.
>
> Measure the distance from this point to the other end of your line.
>
> Do an appropriate addition or subtraction to find out if your measurement is accurate.
>
> Repeat the instructions above using different measurements.
>
> If you can, try working in pairs. One person finds the distance by drawing, the other by doing a subtraction.

> **Learner tip**
>
> When adding or subtracting numbers with units, make you have changed both units so they are the same, e.g. all in cm. Look at the unit on Measurement and standard units for more information on this.

Multiplication and division

How well do you know your multiplication tables? We covered these in Unit 105 and you will find multiplication and division much quicker if you are confident with your tables.

The 7 and 8 multiplication tables can be difficult to remember. Write them down now. 1×7 has been completed for you.

×	1	2	3	4	5	6	7	8	9	10
7	7									
8										

> Remember 5×8 is the same as 8×5

You should also know how to multiply or divide a 2-digit number by a 1-digit number.

For example:

$23 \times 4 = 92$

or

$24 \div 6 = 4$

It will be useful for you to practise these skills while problem solving.

> **Activity**
>
> In pairs, ask each other questions based on multiplication tables.

You should understand that multiplication and division are opposite processes. So if you know that $4 \times 6 = 24$, you should also know that $24 \div 4 = 6$ and $24 \div 6 = 4$.

Make sure you understand the following worked examples.

Example

Work out these multiplication and division sums.

a 32×3
b $96 \div 3$
c 18×7
d $126 \div 7$

Solution

a
$$\begin{array}{r} 3\ 2 \\ \times\quad 3 \\ \hline 9\ 6 \end{array}$$

You multiply first the units and then the tens by 3.
$2 \times 3 = 6$ and $3 \times 3 = 9$

b
$$\begin{array}{r} 3\ 2 \\ 3\overline{)9\ 6} \end{array}$$

You divide first the tens and then the units by 3.
$9 \div 3 = 3$ and $6 \div 3 = 2$

c
$$\begin{array}{r} 1\ 8 \\ \times\quad 7 \\ \hline 1\ 2\ 6 \\ {\scriptstyle 5} \end{array}$$

$8 \times 7 = 56$
You write the 6 in the units column and carry 5 tens over.
$1 \times 7 = 7$; $7 + 5$ carried over $= 12$

d
$$\begin{array}{r} 1\quad 8 \\ 7\overline{)1\ 2\ {}^{5}6} \end{array}$$

7 into 1 does not go, so you look at the next digit. 7 into 12 is 1 remainder 5.
Look at the remainder together with the next digit. 7 into 56 is 8.

Written methods for multiplication with more than one digit

Now you can multiply by one digit, you need to be able to do questions such as 53×38 or 258×63 without a calculator.

There are several methods to do this. Two are shown here but your teacher may show you more. Choose a method you are happy with and stick with it.

Method 1: Long multiplication

```
      5 3
 ×    3 8
 1 5 9 0      ← │ 53 × 30 │
   4 2 4      ← │ 53 × 8  │
     2
 2 0 1 4      ← │ Add     │
 1   1
```

This is the traditional method, called 'long multiplication'.

Method 2: Grid method

The second method uses a grid.

53×38

×	50	3
30	1500	90
8	400	24

```
  1 5 0 0
    4 0 0
      9 0
 +    2 4
  2 0 1 4
  1   1
```

258×63

×	200	50	8
60	12000	3000	480
3	600	150	24

```
  1 2 0 0 0
    3 0 0 0
      4 8 0
      6 0 0
      1 5 0
 +      2 4
  1 6 2 5 4
    1   1
```

Method 3: Lattice method

In this method, a lattice is drawn to help us work out the sum.

Let's try a worked example: 263×47.

Remember to add along the diagonal (red) arrows.

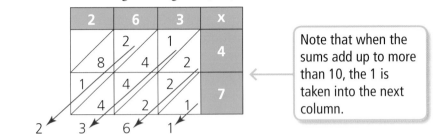

> Note that when the sums add up to more than 10, the 1 is taken into the next column.

Written methods for division with more than one digit

In Stage 1 we looked at two methods for division: long division and chunking. You can use either of the methods, but the solution given just shows the method of long division.

For $840 \div 20$:

```
        0   4   2
   20 | 8   4   0
        8   0
        ─────────
            4   0
            4   0
            ─────────
                0
```

There are no 20s in 8, so look at 84. There are 4 20s in 84.

$4 \times 20 = 80$, so put 4 on the answer line and take 80 away from 84 – remainder 4.

Bring down the zero. How many 20s in 40? 2 with nothing left over.

So $840 \div 20 = 42$

Remainders when dividing

Sometimes when you do a division you will not get a simple answer – in some cases, there will be something left, a remainder.

Let's say, for example, you have a 200 cm length of wood and you need to cut 60 cm lengths from it for a kitchen fitting. How many lengths would you get and what would you have left?

When you divide 200 by 60 you get 3 whole lengths from 180 cm and so you have 20 cm left. There is no need to use decimals in this instance. The question asks for lengths of 60 cm so we do not need to continue dividing.

> ### Example
> What if you have a length of wood of 420 cm? How many lengths of 45 cm can be cut off?
>
> ### Solution
> You might have noticed that 10×45 is 450, which is too much. So if we try 9×45 we get 405, which leaves 15 cm. You can therefore cut 9 lengths and will have 15 cm left over. The question asks for lengths of 45 cm so we do not need to continue dividing.

Sometimes you should continue to divide. This is usually with money or metres where it is common to have decimal places.

> ### Example
> What is $78 shared between 12 people?
>
> ### Solution
> This is $6 with $6 left over. We can continue to divide ($6 \div 12$ doesn't go, so bring down a 0).
> $60 \div 12 = 5$
> $78 \div 12 = 6.5$
> so $\$78 \div 12 = \$6.50$

Tasks

1 Mentally work out how many you need to add to the following numbers to make 100.

   a 55

   b 32

   c 74

2

     1 6 7 2

   +   7 1 9

3 Calculate $1895 + 2314 + 26$

4

     1 6 7 8

   −   7 1 3

5 Calculate $43987 - 21803$

6 Calculate 478×21

7 Calculate 2034×402

8 Calculate $374 \div 17$

9 Calculate $909 \div 30$

10 Calculate $3239 + 106 - 98$

Test your knowledge

1 Niko is selling raffle tickets.
He had 750 tickets to begin with and has already sold 387.
How many tickets does he have left?

  a 363

  b 373

  c 437

  d 473

2 30 765 + 2 192 =

  a 32 857

  b 32 957

  c 32 867

  d 52 685

3 226 × 34 =

  a 1582

  b 7564

  c 7684

  d 9718

4 1023 × 42 =

  a 24 542

  b 42 866

  c 42 966

  d 47 166

5 3240 ÷ 20 =

  a 102

  b 110

  c 112

  d 162

6 141 ÷ 6 =

  a 23

  b 23.5

  c 27.1

  d 271

7 A man wins one million dollars and gives $55 000 to charity.
How much money does he have left?

a $45 000

b $945 000

c $955 000

d $450 000

8 A joiner needs 4 pieces of wood each 25 cm long and another piece
75 cm long.
How much wood does she need in total?

a 100 cm

b 175 cm

c 325 cm

d 400 cm

9 A cook has 1 kg of flour.
He uses 600 g of flour in one recipe and 250 g in another.
How much flour does he have left?

a 850 g

b 400 g

c 250 g

d 150 g

10 How many sheets of paper are needed to print twenty-five copies of a
225 page report, using one sheet per page?

a 9

b 1575

c 5405

d 5625

Unit 204
Operations on decimal fractions

Introduction

Mathematics has three ways to describe parts of numbers: using a fraction, a decimal or a percentage. This unit will cover addition, subtraction, multiplication and division of decimal fractions.

This unit builds on the Stage 1 topic on decimal fractions (Unit 106). Before tackling this topic you may like to remind yourself of what a fraction means and how decimal place values work.

This unit also builds on the skills of addition, subtraction, multiplication and division you covered in an earlier unit from this stage, Unit 203, looking at whole numbers. As you cover new topics in mathematics, you will find the ideas often build on something you have covered before so make sure you understand each topic and ask for help if you are unsure.

Learning objectives

In this unit you will:

- add and subtract two numbers with no more than three decimal places
- multiply a number with up to four digits and no more than two decimal places by a whole number with up to two digits
- divide a number with not more than four digits and two decimal places by a whole number with up to two digits
- solve problems using a combination of operations.

You will find decimal fractions useful when working with measures in the decimal system: kilograms (for weight), metres (for distance and length) and litres (for capacity) are all based on tens.

Addition and subtraction of decimal fractions

You can add and subtract decimals in the same way as whole numbers. There are various different methods you can use. Choose the method, or methods, you find easiest to understand.

Column addition and subtraction

Make sure you line up the decimal points and the tenths and hundredths and thousandths after it as well as the hundreds, tens and units before it.

Example 1

Use column addition to work out 16.45 + 2.62.

Solution

```
    1 6 . 4 5
  +  2 . 6 2
    1 9 . 0 7
       1
```

First, set the addition in columns.

Add the digits in the hundredths column (5 + 2 = 7, write 7 in the hundredths column).

Add the digits in the tenths column (4 + 6 = 10, write the 0 in the tenths column and write the 1 under the units).

Add the digits in the units column (6 + 2 = 8, 8 + 1 = 9, write the 9 in the units column).

There is only 1 digit in the tens column, so write this (1).

16.45 + 2.62 = 19.07

Example 2

Use column subtraction to work out 13.78 − 1.24.

Solution

```
    1 3 . 7 8
  −  1 . 2 4
    1 2 . 5 4
```

First, set the subtraction in columns.

Subtract the digits in the hundredths column (8 − 4 = 4, write 4 in the hundredths column).

Subtract the digits in the tenths column (7 − 2 = 5, write the 5 in the tenths column).

Subtract the digits in the units column (3 − 1 = 2, write the 2 in the units column).

There is only 1 digit in the tens column, so write this (1).

13.78 − 1.24 = 12.54

Example 3

Work out 124.5 + 79.87.

Solution

First, set the addition in columns

```
    1 2 4 . 5
  +   7 9 . 8 7
    2 0 4 . 3 7
      1   1   1
```

Add the digits in the hundredths column (the hundredths column in the top number is empty, which is the same as 0, 0 + 7 = 7, write 7 in the hundredths column).

Add the digits in the tenths column (8 + 5 = 13, write the 3 in the tenths column and write the 1 under the units).

Add the digits in the units column (4 + 9 = 13, 13 + 1 = 14, write the 4 in the units column and write the 1 under the tens).

Add the digits in the tens column (2 + 7 = 9, 9 + 1 = 10, write the 0 in the tens column and write the 1 under the hundreds).

Add the digits in the hundreds column (1 + 1 = 2, write the 2 in the hundreds column).

124.5 + 79.87 = 204.37

You cannot take 8 hundredths from 0 hundredths so you will need to use partitioning to split the number differently. Instead of 2 tens + 3 units + 4 tenths, 23.4 becomes 2 tens + 3 units + 3 tenths + 10 hundredths. Change the 4 hundredths to 3 hundredths to show that you have moved 1 tenth to the hundredths column and add in 10 hundredths. Now you can subtract the hundredths (10 − 8 = 2).

You cannot take 5 tenths from 3 tenths so you will need to use partitioning to split the number differently. Instead of 2 tens + 3 units + 3 tenths, 23.3 becomes 2 tens + 2 units + 13 tenths. Change the 3 units to 2 units to show that you have moved 1 unit to the tenths column and add in 10 tenths. Now you can subtract the tenths (13 − 5 = 8).

You cannot take 9 units from 2 units so use partitioning: instead of 20 + 2, 22 becomes 10 + 12. Change the 2 tens to 1 ten to show that you have moved 1 ten to the units column and add in 10 units. Now you can subtract the units (12 − 9 = 3).

Example 4

Work out 23.4 − 9.58.

Solution

First set the subtraction in columns

```
  ¹2 ¹²3 . ¹³4 ¹0
  −     9 . 5 8
    1 3 . 8 2
```

Subtract the digits in the hundredths column. (The hundredths column in the top number is empty, which is the same as 0.) Write 2 in the hundredths column.

Subtract the digits in the tenths column. Write 8 in the tenths column.

Subtract the digits in the units column. Write 3 in the units column.

There is only 1 digit in the tens column, so write this (1).

23.4 − 9.58 = 13.82

Partitioning with decimals

Example 1

Work out 17.3 − 8.9.

Solution

Split the second number into integer part plus decimal part
(8.9 = 8 + 0.9).

Subtract the parts in turn (17.3 − 8 = 9.3; 9.3 − 0.9 = 8.4).

You can break 0.9 down as 0.3 + 0.6 and subtract it in two parts: 9.3 − 0.3
= 9, 9 − 0.6 = 8.4.

Example 2

Work out 12.67 + 4.55.

Solution

Add the decimal parts (0.67 + 0.55 = 1.22) using any appropriate
method,

e.g. 0.67 + 0.55 = 0.7 + 0.55 − 0.03 = 1.25 − 0.03 = 1.22

Add the integer parts, i.e. the parts of the number that appear before the
decimal point

12 + 4 = 16

Add the results from both steps: 16 + 1.22 = 17.22

Complements with decimals

To work out the difference between 17.3 and 8.9:

Find a complement to 10 of 8.9, which is 1.1

Complete the equivalent 2-part calculation: 17.3 − 10 = 7.3, 7.3 + 1.1 = 8.4.

Counting on with decimals

Draw or use a mental image of a number line to count on from 8.9.

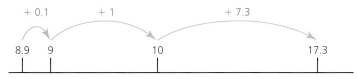

You count on 0.1, 1 and 7.3.

So add these (0.1 + 1 + 7.3) = 8.4

Rounding

Subtract 9 from 17.3 to get 8.3

You have taken away 0.1 too many, so compensate by adding 0.1 to 8.3 and
the answer is 8.4

Activity

Use your ruler to draw a line
which is 5.6 cm long.

Mark on the line a point which is
2.8 cm from one end.

Measure the distance from this
point to the other end of your
line.

Do an appropriate addition or
subtraction to find out if your
measurement is accurate.

Multiplication and division of decimal fractions

Multiplying a decimal by an integer

Compare these two multiplications for finding the cost of three CDs at $4.90 each.

Working in cents

You know that $4.90 = 490 cents.

```
    4 9 0
×       3
  1 4 7 0 = $14.70
```

Working in dollars

```
    4 . 9 0
×         3
$ 1 4 . 7 0
```

Now try these decimal multiplication calculations.

> **Example 1**
>
> 41.25 × 13
>
> Solution
>
> Remember to set out the sum.
> ```
>         41.25
> ×          13
>        123.75
>        412.50
>        536.25
> ```

> **Example 2**
>
> 0.382 × 72
>
> Solution
> ```
>         0.382
> ×          72
>         0.764
>        26.740
>        27.504
> ```

Multiply first the units, then the tens and then the hundreds by 3.

Write the units under the units, the tens under the tens, and so on.

Convert your answer from cents to dollars by dividing by 100.

To multiply a decimal by an integer, put the decimal points under each other.

Make sure you line up your work carefully.

Put the first digit you work out under the last decimal place.

In each case, the digits are the same.

Learner tip

Make a rough estimate of the answer to questions like these examples.

41 × 10 = 410 so the answer will not be 53.625 or 5362.5

Similarly, 0.382 is less than one so the answer will be less than 72.

You can also count the number of digits after the decimal point in the question and check there is the same number of digits after the decimal point in the answer.

Dividing a decimal by an integer

Working in cents

You know that $320.80 ÷ 40 is the same as 32 080 cents shared by forty.

For 32 080 ÷ 4 this is just one method you can use. See Unit 105 on multiplication and division of whole numbers in Stage 1.

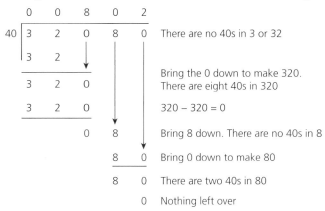

	0	0	8	0	2	
40	3	2	0	8	0	There are no 40s in 3 or 32

Bring the 0 down to make 320.
There are eight 40s in 320

320 − 320 = 0

Bring 8 down. There are no 40s in 8

Bring 0 down to make 80

There are two 40s in 80

Nothing left over

So 32 080 ÷ 40 = 802

Convert your answer from cents to dollars by dividing by 100.

802 ÷ 100 = 8 dollars and 2 cents

Working in dollars

$$40\overline{)3\,2\,0\,.\,8\,0}^{\;\;\;8\,.\,0\,2}$$

Complete the calculation as before.

Remember, line up the decimal point on the answer line above the decimal point in the number before you start to divide.

> ### Example 1
>
> 1.95 ÷ 13
>
> ### Solution
>
> 13 does not go into 1, try 19 – goes once with 6 left over.
>
> Bring down the 5 so 13 into 65.
>
> 5 × 13 = 65
>
> The answer is 0.15

Example 2

7202.70 ÷ 30

Solution

30 does not go into 7, try 72 – goes twice with 12 left over.

Bring down the 0 so 30 into 120.

4 × 30 = 120.

30 does not go into 2 or 27. (Remember to put the zeros on the answer line.)

Try 270.

9 × 30 = 270

The answer is 240.09

Tasks

1 Add 2.75 and 13.025

2 13.05 + 2.345

3 16 − 0.755

4 1706.5 − 32.9

5 13.25 × 29

6 107.8 × 52

7 247.8 ÷ 2

8 0.249 ÷ 30

9 3.27 m + 0.5 m + 2.66 m

10 An item costs $250 cash or you can pay $25 deposit and 12 monthly payments of $27.50.
What is the total cost for the second method?
What is the difference between paying cash or paying in installments?

Test your knowledge

1 A bag of red apples weighs 1.5 kg and a bag of green apples weighs 2.9 kg. What is the total weight?

   a 3.4 kg   **c** 30.5 kg

   b 4.4 kg   **d** 44 kg

2 What is 167.34 + 10.7?

   a 168.41   **c** 177.41

   b 177.04   **d** 178.04

3 What is 21 − 2.735?

   a 6.35   **c** 19.265

   b 18.265   **d** 19.735

4 What is 14.05 × 12?

   a 42.15   **c** 169.80

   b 168.60   **d** 174.00

5 What is 36.28 × 34?

   a 1111.22   **c** 1333.52

   b 1233.52   **d** 1560.04

6 What is 30.24 ÷ 8?

   a 3.03   **c** 3.78

   b 3.3   **d** 3.8

7 What is 10.5 ÷ 21?

   a 0.5   **c** 5

   b 2   **d** 11

8 What is 2.56 m + 2.6 m − 3.2 m?

   a 3.16   **c** 1.96

   b 2.42   **d** 1.16

9 What is 3.84 × 10 ÷ 100?

   a 0.384   **c** 38.4

   b 3.84   **d** 384

10 What is 49.4 − 6.56?

   a 42.84   **c** 43.94

   b 43.16   **d** 43.96

Unit 205
Operations on common fractions

Introduction

A fraction is a way of describing a number which is less than one whole. At Stage 1, in Unit 107, we worked with the simple fractions: one-half $\left(\frac{1}{2}\right)$ and one-quarter $\left(\frac{1}{4}\right)$. Remember, in a fraction, the top number is called the **numerator** and tells you how many parts the fraction shows. The bottom number is called the **denominator** and tells you how many equal parts the whole is split into.

Learning objectives

In this unit you will find information to:

- develop this knowledge and add and subtract other fractions
- calculate a common fraction of a quantity and use equivalent fractions.

How to find a simple fraction of an amount

If you want to find half of a number, you can either multiply it by one-half or divide it by two. Both ways will give you the same answer.

So, half of 24 is 12 as shown below:

$24 \times \frac{1}{2} = 12$

or

$24 \div 2 = 12$

> You can do this for any fraction which has '1' as the numerator (on the top).

So, a third of 27 is either $27 \times \frac{1}{3} = 9$ or $27 \div 3 = 9$

> **Activity**
>
> In pairs, use the fractions $\frac{1}{4}$ and $\frac{1}{10}$ to find a fraction of an amount.

> **Example**
>
> A jacket has a price tag showing $75. In a sale, all ticket prices are reduced by one-third.
>
> How much is the discount off the jacket price in the sale?
>
> **Solution**
>
> $75 \div 3 = 25$
>
> The discount is $25.

Equivalences

Let's consider this diagram of a pizza.

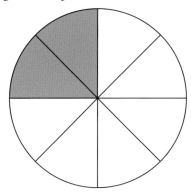

In the image, two-eighths $\frac{2}{8}$ of the pizza is shaded (two parts out of the total eight).

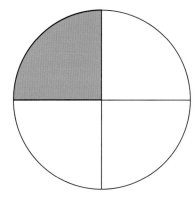

In this second image, one-quarter $\left(\frac{1}{4}\right)$ of the pizza is shaded. As you can clearly see, the same amount of the pizza is shaded in each picture. So two-eighths and one-quarter are equivalent fractions.

Now use the next image of a pizza to work out the equivalencies for one half in quarters and in eighths.

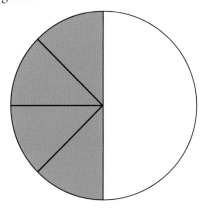

You can also do this by calculation. $\frac{1}{4} = \frac{?}{8}$

The proportion of a whole remains the same if you multiply both the numerator and the denominator by the same number. We know that we need to multiply 4 by 2 to get the answer 8 so we multiply the numerator by 2.

$1 \times 2 = 2$

Therefore, $\frac{1}{4} = \frac{2}{8}$

Activity

Can you find the equivalence of $\frac{1}{5}$ in tenths?

You may find it easier to think of a chocolate bar rather than a pizza but the answer will be the same.

Now try this by calculation.

Adding fractions

To add fractions all the fractions must be the same size such as all halves or all quarters.

$\frac{1}{2} + \frac{1}{2} = \frac{2}{2}$

Think of the pizza. If you have two halves, you have a whole. So one-half and one-half gives you two halves, which is the same as a whole.

We can add quarters in the same way

$\frac{1}{4} + \frac{1}{4} + \frac{1}{4} + \frac{1}{4} = \frac{4}{4}$

Think of the pizza again. If you have four-quarters, you have a whole. So one-quarter add one-quarter add one-quarter add one-quarter gives you four-quarters, which is the same as a whole.

How can we add one-half and one-quarter?

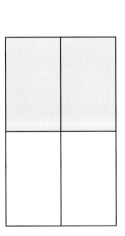

We need to change the half to quarters. Think of the pizza or this piece of paper.

We can see that a whole is 4 quarters or 2 halves. We can also see that $\frac{1}{2}$ is the same as two quarters or $\frac{2}{4}$.

If we add two-quarters and one-quarter, we get three-quarters.

$\frac{2}{4} + \frac{1}{4} = \frac{3}{4}$

We can add these across the numerators (the top number) because this tells us how many parts we have. But we can only do this if the denominators (the bottom number) are the same. Note we do not add the bottom numbers.

We can also write this as $\frac{2 + 1 = 3}{4}$

What is $\frac{3}{4} + \frac{1}{4}$? $\frac{3+1=4}{4}$

$\frac{4}{4}$ is one whole so

$\frac{3}{4} + \frac{1}{4} = 1$

What is $\frac{3}{5} + \frac{3}{10}$?

Here the denominator is not the same so we cannot just add the top numbers.

Think of the equivalent fractions section you have just worked through. How can we get the denominators the same?

$\frac{1}{5} = \frac{2}{10}$ (look through the equivalencies section again, if you need to).

If $\frac{1}{5}$ is $\frac{2}{10}$, then $\frac{3}{5}$ must be 3 lots of $\frac{2}{10}$, so $\frac{2}{10} + \frac{2}{10} + \frac{2}{10}$ which is $\frac{6}{10}$.

Now we can add, as the denominators are the same:

$\frac{6}{10} + \frac{3}{10} = \frac{6+3}{10} = \frac{9}{10}$

Try these.

1 $\frac{7}{10} + \frac{3}{100}$

2 $\frac{3}{4} + \frac{2}{8}$

3 $2\frac{1}{2} + \frac{7}{10}$

Subtracting fractions

To subtract fractions, all the fractions must be the same size such as all halves or all quarters.

$\frac{3}{4} - \frac{1}{4} = \frac{2}{4}$

Think of the pizza. If there are three-quarters on a plate and you take one-quarter, there are two-quarters left on the plate. We can say there are two-quarters left but we usually call two-quarters one-half.

Draw a pizza to check this if you are not certain.

What is $\frac{1}{2} - \frac{1}{4}$?

To subtract fractions all the fractions must be the same size. A quarter is not as big as a half so we need to make the half into quarters.

$\frac{1}{2} = \frac{2}{4}$

Now we can do the subtraction sum

$\frac{2}{4} - \frac{1}{4} = \frac{1}{4}$

Think of the pizza again. If you had a pizza cut into 4 and half of it was on the plate, there would be two-quarters on the plate. You take away one-quarter and one-quarter is left.

You often need to subtract fractions from a whole.

You have 3 litres of oil and use half a litre. How much oil is left?

Think of this as pizzas if it helps. You can change 3 into 6 halves.

$\frac{6}{2} - \frac{1}{2} = \frac{5}{2}$

Think of the pizzas. How many whole pizzas and halves are left?

2 whole and one half = $2\frac{1}{2}$

Therefore, there are $2\frac{1}{2}$ litres of oil left.

Alternatively, you can just change one litre into 2 halves and remember you have 2 full litres. This is similar to what you did with hundreds, tens and units in the earlier chapters.

$\frac{2}{2} - \frac{1}{2} = \frac{1}{2} + 2$ whole litres $= 2\frac{1}{2}$ litres.

Try these. Use whichever method you find easier.

1 $3 - \frac{3}{5}$

2 $\frac{1}{2} - \frac{3}{8}$

3 $\frac{1}{10} - \frac{1}{100}$

Tasks

1 Find one-third of $66.

2 What is $\frac{2}{10}$ of 230?

3 Match the equivalent fractions.

$\frac{1}{2}$	$\frac{10}{100}$
$\frac{1}{10}$	$\frac{2}{8}$
$\frac{1}{4}$	$\frac{5}{10}$

4 $\frac{1}{3} = \frac{?}{6}$

5 $\frac{1}{2} + \frac{3}{10}$

6 $1\frac{1}{4} + \frac{5}{8}$

7 $5 - \frac{3}{10}$

8 $\frac{3}{4} - \frac{5}{8}$

9 $1\frac{1}{2} - \frac{3}{8}$

10 What is the total weight of a shopping bag which contains 2 kg of potatoes, 1 kg of apples, half a kilogram of tomatoes and $\frac{3}{4}$ kilogram of grapes?

Test your knowledge

1 What is a quarter of 24 m?

  a 4 m

  b 6 m

  c 12 m

  d 14 m

2 Find $\frac{1}{5}$ of 45.

  a 5

  b 6

  c 9

  d 15

3 Which of these are equivalent fractions?

  A $\frac{1}{4}$

  B $\frac{3}{6}$

  C $\frac{3}{4}$

  D $\frac{4}{8}$

  a A and C

  b B and C

  c C and D

  d B and D

4 $\frac{3}{5} + \frac{3}{10} =$

  a $\frac{6}{15}$

  b $\frac{6}{5}$

  c $\frac{6}{10}$

  d $\frac{9}{10}$

5 $2\frac{3}{8} + 1\frac{1}{4} =$

  a $3\frac{4}{4}$

  b $3\frac{4}{8}$

  c $3\frac{4}{12}$

  d $3\frac{5}{8}$

6 $\frac{5}{8} - \frac{1}{4} =$

a $\frac{4}{4}$

b $\frac{3}{8}$

c $\frac{4}{8}$

d $\frac{4}{12}$

7 $3\frac{3}{5} - \frac{4}{10} =$

a $2\frac{2}{10}$

b $\frac{11}{10}$

c $3\frac{1}{5}$

d $\frac{11}{5}$

8 A shop offers $\frac{1}{3}$ off all prices.

An item costs $180.
What is one-third of $180?

a $30

b $33

c $45

d $60

9 What is $\frac{3}{10} + \frac{5}{10} + \frac{1}{100}$?

a $\frac{9}{10}$

b $\frac{9}{120}$

c $\frac{36}{100}$

d $\frac{81}{100}$

10 A piece of wood is 3 m long.

A man cuts off one piece $\frac{1}{2}$ m and another piece 1 m long.
How much wood is left?

a $\frac{1}{2}$ m

b $1\frac{1}{2}$ m

c 2 m

d $2\frac{1}{2}$ m

Introduction

In the last unit we looked at fractions as a way of describing a number which is less than one whole. Percentages are another way to show and compare parts of a number.

Percent means 'out of 100'. Think of 100 cents in a dollar.

We use the symbol % to show a percentage.

Learning objectives

In this unit you will find information on:

- expressing numerical information as a percentage
- calculating percentages of numbers.

This will help you to prepare for questions about:

- finding percentage discounts
- finding percentage interest
- expressing one number as a percentage of another.

Expressing simple numerical information as a percentage

Percentages are parts of a whole. 50% means 50 parts in 100 or 50 out of 100. This is the same as the fraction $\frac{50}{100}$. Percentages are always out of 100.

You can find the percentage of a number by finding the fraction of 100.

If a box contains 100 pens and 3 pens are black, the percentage of black pens is $\frac{3}{100}$ or 3%.

If the box contains 100 pens and 65 are blue, the percentage of blue pens is $\frac{65}{100}$ or 65%.

The remaining pens in the box are red. What percentage of the pens are red? ←

> You can add up the number of blue pens and black pens and take that total from 100, so
>
> 3 + 65 = 68, 100 − 68 = 32. This is 32 out of 100, $\frac{32}{100}$ or 32%.
>
> Or, add 3% and 65% = 68%, then 100 − 68 = 32%.

There are 10 questions in a test. A student gets 6 questions correct. What percentage is this? Percentages must be out of 100.

$$\frac{6}{10} = \frac{60}{100}$$

What is 3 out of 4 as a percentage?

This is the same as $\frac{3}{4}$.

What is $\frac{1}{4}$ of 100? It is 25.

We need to find $\frac{3}{4}$ so $3 \times 25 = 75$.

$\frac{3}{4} = \frac{75}{100}$, so this is 75%.

> Look again at the last unit on fractions, if you are unsure of this.
> $\frac{60}{100} = 60\%$

Finding percentage parts of numbers

Percentages are parts of a whole. 50% means 50 parts in 100 or 50 out of 100. This is the same as the fraction $\frac{50}{100}$. Percentages are always out of 100.

You may see a sign in a sale which says '10% off today'.

What does this mean?

It means 10 out of 100. So if the item costs $100, there is $10 off today.

If the item costs $1 this is 100 cents, so there is 10 cents off today.

If the item costs $20, you can think of this as 20×10 cents = 200 cents or $2.

Or, you can find 1%.

1% is $\frac{1}{100}$ of 20 or $\frac{20}{100}$ which is 0.2.

If 1% is 0.2, then 10% is $10 \times 0.2 = \$2$.

> Look again at the unit on decimals if you are unsure of this.

Example

a What is 10% of 500?

b What is 25% of 500?

c What is 70% of 500?

Solution

a 1% is $\frac{500}{100} = 5$

10% is $10 \times 5 = 50$

b 1% is $\frac{500}{100} = 5$

25% is $25 \times 5 = 125$

c 1% is $\frac{500}{100} = 5$

70% is $70 \times 5 = 350$

Try finding 30% of \$60

There are different methods you can use.

Method 1 ←

Use whichever method you find easier.

1% is $\dfrac{60}{100} = 0.6$

30% is $30 \times 0.6 = \$18$

Method 2

1% is $\dfrac{60}{100} = 0.6 = 60$ cents

30% is 30×60 cents $= 1800$ cents. $\dfrac{1800}{100} = \$18$

Method 3: Using fractions

$30\% = \dfrac{30}{100} = \dfrac{3}{10}$

Then you calculate $\dfrac{3 \times 60}{10 \times 1} = \dfrac{180}{10} = \$18$

Example

A travel agent offers a 10% reduction for holidays arranged on the internet. The normal cost of a holiday is \$360.

What is the cost of the holiday?

Solution

10% of 360 is $\dfrac{10}{100} = 36$.

The reduction is \$36.

So the holiday costs $360 - 36 = 324$ or \$324.

Activity

Can you see an easy way to find 10% of an amount?

How could this help you find 20% or 30%?

10% is the same as dividing by 10. 20% is not the same as dividing by 20 but it is the same as two lots of 10%.

Tasks

1 Find 30% of 200.

2 A shop has 25% off in a sale. What is the discount off a pair of shoes costing \$68?

3 Find 5% of 420.

4 What is 3 out of 10 expressed as a percentage?

5 What is 4 out of 20 expressed as a percentage?

6 In a group of 50 people, two do not eat meat. What percentage of the people in the group does not eat meat?

7 A saver receives 2% interest on his savings. How much does he receive on savings of \$4000?

8 What is one out of five as a percentage?

9 Find 80% of 50.

10 What is 6 out of 25 expressed as a percentage?

1 A quality control department checks 100 items and finds 4 have faults.
What is this expressed as a percentage?

a 4%

b 25%

c 40%

d 96%

2 Twenty students are given homework but only 17 hand their homework in on time.
What percentage of the students hand their homework in on time?

a 17%

b 30%

c 68%

d 85%

3 A car sales representative sells 50 cars one month.
She wants to sell 10% more next month.
How many more does she want to sell?

a 5

b 10

c 20

d 40

4 Find 15% of 60.

   a 6

   b 9

   c 15

   d 45

5 Find 20% of 120.

   a 5

   b 6

   c 20

   d 24

6 What is 3% of $1000?

   a $0.3

   b $3

   c $30

   d $300

7 What is 5 out of 25 expressed as a percentage?

   a 4%

   b 5%

   c 20%

   d 25%

8 What is 60 out of 200 expressed as a percentage?

   a 3%

   b 12%

   c 30%

   d 33%

9 What is 70% of 300?

   a 43

   b 70

   c 210

   d 230

10 Express 10 out of 1000 as percentage.

   a 0.1%

   b 1%

   c 10%

   d 100%

Introduction

In Unit 101 we introduced decimal fraction and common fraction equivalencies for halves and quarters and in Unit 206 in this stage we looked at percentages as fractions. Common fractions, decimal fractions and percentages are all ways of describing a number which is less than one whole.

Fractions, decimals and percentages are used all the time in everyday life. For example, you see shop signs advertising sales of $\frac{1}{2}$ price or 20% off and it is important that you can compare fractions and percentages so that you can work out which is the best offer.

Learning objectives

In this unit you will find information on:

- recognising equivalencies of common fractions, decimal fractions and percentages
- how to convert common fractions to decimal fractions.

This will help you to prepare for questions about:

- equivalencies and conversions.

Equivalencies of common fractions, decimal fractions and percentages

In Unit 206 we said that 50% means 50 parts in 100 or 50 out of 100 which is the same as the fraction $\frac{50}{100}$.

So 50% = $\frac{50}{100}$.

From the work on place value in Unit 201 earlier in this stage, we know that $\frac{50}{100}$ can be written as 0.50 and this can be written as 0.5 because we do not need the final zero here.

$\frac{5}{10}$ is the same as $\frac{50}{100}$

So 50% = 0.50 = 0.5 = $\frac{50}{100}$ = $\frac{5}{10}$. These values are all equivalent values or **equivalencies**.

From Unit 205 on fractions, we know that $\frac{5}{10} = \frac{1}{2}$.

It is more usual, or common, to call five-tenths 'one-half'.

A common equivalency is $\frac{1}{2}$ = 0.5 = 50%.

Activity

The first line of this table has been completed for you. Can you complete the gaps?

Fraction	Decimal	Percentage
$\frac{1}{2}$	0.5	50%
	0.25	
		20%
	0.1	
$\frac{1}{20}$		

For the fractions, it may still be useful to have a mental image of a pizza being cut up or a bar of chocolate. For the decimals, you may like to think of cents and dollars or centimetres and metres. Remember the percentages are per 100, i.e. hundredths.

Although most people think of fractions, decimals and percentages as very different, they are really the same idea written in different ways.

So, for example, the fraction $\frac{1}{4}$ is the same as 25%, which is the same as the decimal 0.25. For numbers bigger than one we can also use all three.

For example, $1\frac{1}{2}$ is 150%, which is the same as 1.5.

We tend to choose the one that best suits the problem we are solving or the context we are working in.

Which would you be more likely to say?

Was $100
Now $75

That's a 0.25 discount.

That's a $\frac{1}{4}$ discount.

That's a 25% discount.

Most shops would tend to use '25% discount' because it sounds more, even though all are the same amount.

Check you have the correct answers in your table. You will find it useful to learn these equivalencies.

Fraction	Decimal	Percentage
$\frac{1}{2}$	0.5	50%
$\frac{1}{4}$	0.25	25%
$\frac{1}{5}$	0.20	20%
$\frac{1}{10}$	0.1	10%
$\frac{1}{20}$	0.05	5%

Converting common fractions to decimal fractions

A fraction such as $\frac{1}{2}$ means a whole divided by two so a fraction such as $\frac{1}{8}$ means a whole divided by eight. $\frac{3}{8}$ is three of these parts. How can we convert $\frac{3}{8}$ to a decimal fraction? Thinking about fractions you know will help you to work with other fractions.

$$2 \overline{)\,1.0\,}^{\,0.5}$$

> Remember the decimal point in the answer goes above the decimal point in the original value. So the answer is 0.5

2 into 1 does not go, so try 2 into 10 – yes, there are five 2s in 10.

Now try $\frac{3}{8}$. This is 3 divided by eight.

	0.	3	7	5	
8	3.	0	0	0	8 into 3 does not go. Try 8 into 30 – there are three 8's in 30 with remainder 6.
	2	4			
		6	0		Bring down a zero. How many 8s in 60?
		5	6		7 × 8 = 56 remainder 4
			4	0	Bring down a zero. How many 8s in 40? 5 × 8 = 40.

> If you are unsure of this, look again at Unit 4.

$\frac{3}{8}$ is 0.375 as a decimal. Or, as a percentage, 37.5%.

Activity

Use two different methods of working out $\frac{3}{20}$ as a decimal.

You could use a division sum for one way and information from the table as the other way.

Learner tip

Remember how to write the division calculation. $\frac{3}{8}$ is three divided by 8, not 8 divided by 3.

Which is the longest distance, half a kilometre or 0.45 kilometres?

Change both values to fractions or both to decimals. Remember if you want to compare fractions they both need to have the same denominator.

$\frac{1}{2}$ is 0.5. Compare 0.5 with 0.45.

0.5 is more than 0.45, so $\frac{1}{2}$ km is further than 0.45 km.

Or, $\frac{1}{2}$ is $\frac{50}{100}$ and 0.45 is $\frac{45}{100}$.

$\frac{50}{100}$ is more than $\frac{45}{100}$, so $\frac{1}{2}$ km is further than 0.45 km.

Tasks

Copy and complete the boxes in questions 1–7 using these numbers:

$\frac{3}{4}$ $\qquad$ $\frac{1}{10}$ $\qquad$ 100% $\quad$ 0.3 $\qquad$ 25% $\quad$ 0.1 $\qquad$ 75%

0.5 $\qquad$ $\frac{1}{5}$ $\qquad$ 20% $\qquad$ $\frac{1}{4}$ $\qquad$ 1 $\qquad$ 50% $\qquad$ $\frac{3}{10}$

	Decimal	Fraction	Percentage
1	1		
2	0.75		
3		$\frac{1}{2}$	
4			30%
5	0.25		
6	0.2		
7			10%

8 What is $\frac{1}{8}$ as a decimal fraction?

9 An item costs $200.

  Three shops sell the item with different discounts. What is the largest discount?

  $\frac{1}{4}$ off $\qquad$ 40% off $\qquad$ $40 off

10 Put these fractions in order of size, smallest first.

  0.2 $\qquad$ $\frac{1}{2}$ $\qquad$ 0.12 $\qquad$ $\frac{5}{100}$

1 What is 0.05 as a fraction?

   a $\frac{5}{1}$

   b $\frac{5}{10}$

   c $\frac{5}{100}$

   d $\frac{5}{1000}$

2 What is a quarter as a percentage?

   a 4%

   b 14%

   c 20%

   d 25%

3 What is 40% as a fraction?

   a $\frac{40}{10}$

   b $\frac{2}{5}$

   c $\frac{4}{5}$

   d $\frac{1}{4}$

4 An item costs $60. Which is the largest discount?

   a $\frac{1}{3}$ off

   b 30% off

   c $30 off

   d $\frac{10}{100}$ off

5 What is 0.7 as a percentage?

   a 0.7%

   b 7%

   c 17%

   d 70%

6 Which is the longest length of wood?

   a $\frac{1}{2}$ m

   b 0.2 m

   c 0.02 m

   d $\frac{1}{10}$ m

7 Which two of the following are equivalent?

A $\frac{1}{5}$

B 0.5

C $\frac{5}{100}$

D 5%

a A and B

b B and C

c C and D

d A and D

8 Which two of the following are equivalent?

A 20%

B $\frac{2}{10}$

C 0.02

D $\frac{1}{2}$

a A and B

b B and C

c C and D

d A and D

9 Which two of the following are equivalent?

A 4%

B 0.4

C 0.04

D $\frac{1}{4}$

a A and B

b A and C

c B and C

d B and D

10 Which two of the following are equivalent?

A 80%

B $\frac{1}{8}$

C 0.08

D $\frac{4}{5}$

a A and B

b B and C

c C and D

d A and D

Unit 208
Orders of magnitude

Introduction

Sometimes you only need to use an approximate number, such as counting large groups of people or making sure you have enough money to pay a bill. When you are paying for an item you may not have the exact notes and coins you need so you offer a convenient overpayment and the shop assistant will give you the required change. Sometimes we need a number to be exact and sometimes a close amount will do.

Learning objectives

In this unit you will find information on:

- building on your understanding of place value (Unit 201) including working with decimal fractions
- rounding numbers to the nearest whole number, ten, hundred or thousand
- rounding decimal fractions to one, two or three decimal places.

Rounding numbers to the nearest ten, hundred and thousand

For large numbers it is usual to approximate to the nearest hundred or thousand. With numbers this size, an exact number may not matter.

A report says that a new concert venue has an attendance of 79 000 on the first evening.

This does not mean that actually 79 000 people were at the concert that evening, it's just easier to report.

The number of people that were actually there was 78 632.

Counting in thousands, 78 632 is between 78 000 and 79 000.

It is nearer to 79 000, as you can see on this number line.

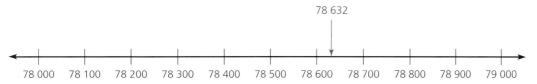

So 78 632 to the nearest thousand is 79 000.

Here is a quick method for rounding to the nearest thousand.

Step 1: put a ring round the thousands digit. For example, 78632.

Step 2: look at the next digit to the right.

If it is less than 5, leave the thousands digit as it is.

If it is 5 or more, add 1 to the thousands digit.

Step 3: replace the remaining digits by zeros.

For the example above, 79 000.

A similar method can be used to round to the nearest 100:

Step 1: put a ring round the hundreds digit. For example, 78632.

Step 2: look at the next digit to the right.

If it is less than 5, leave the hundreds digit as it is.

If it is 5 or more, add 1 to the hundreds digit.

Step 3: replace the remaining digits by zeros.

For the example above, 78600.

Or to the nearest 10:

Step 1: put a ring round the tens digit. For example, 78632.

Step 2: look at the next digit to the right.

If it is less than 5, leave the tens digit as it is.

If it is 5 or more, add 1 to the tens digit.

Step 3: replace the remaining digits by zeros.

For the example above, 78630.

Rounding numbers to the nearest whole number

Sometimes we need to round to the nearest whole number.

Look at this number line.

1.7 is nearer to 2 than 1. So 1.7 to the nearest whole number is 2.

2.4 is nearer to 2 than 3. So 2.4 to the nearest whole number is 2.

Look at this number line.

6.5 is halfway between 6 and 7.

You always round up any numbers that are in the middle.

So 6.5 to the nearest whole number is 7.

To round to the nearest whole number, look at the first decimal place.

- If it is less than 5, leave the whole number as it is.
- If it is 5 or more add 1 to the whole number.

You ignore any digits in the second decimal place and further to the right.

Rounding to a given number of decimal places

Rounding to a given number of decimal places is often used in everyday situations.

When a number is written in decimal form, the digits on the right-hand side of the decimal point are known as **decimal places**.

Numbers can have many different decimal places.

65.3 is written to 1 decimal place.

25.27 is written to 2 decimal places.

98.654 is written to 3 decimal places.

And so on.

You can shorten 'decimal place' to 'd.p.'.

When working with numbers, you may be asked to round a number to a certain number of decimal places. You can adapt the method you learned for rounding earlier to do this.

- Count the decimal places from the decimal point and look at the first digit you need to remove.
- If this digit is less than 5, just remove all the unwanted places.
- If this digit is 5 or larger, add 1 to the digit in the last decimal place you want and then remove the unwanted decimal places.

Example 1

Round 2.658 97 to 3 decimal places.

Solution

8 is the third decimal place so look at the next digit to the right.

The digit is more than 5 so add 1 to the digit in the last decimal place you want and then remove the unwanted decimal places.

2.65897 is 2.659 written to 3 d.p.

Example 2

Round 88.653 to 1 decimal place.

Solution

The second decimal place is 5 so you add 1 to the digit in the first decimal place.

88.653 to 1 decimal place is 88.7 (to 1 d.p.)

Example 3

Round 327.556 to 2 decimal places.

Solution

The third decimal place is larger than 5 so you add 1 to the digit in the second decimal place.

327.556 to 2 decimal places is 327.56 (to 2 d.p.)

Tasks

1 Round 45 240 to the nearest 100.

2 Round 458 900 to the nearest 1 000.

3 Round 6 375 to the nearest 10.

4 Round 4.91 to the nearest whole number.

5 Round 17.32 to the nearest whole number.

6 Round 91.5 to the nearest whole number.

7 Round 4.032 to the nearest whole number.

8 Round 146.9 to the nearest whole number.

9 Round 65.533 to 1 decimal place.

10 Round 21.334 to 2 decimal places.

Test your knowledge

1 Round 6.76 to the nearest whole number.

  a 6.7   **c** 6

  b 6.8   **d** 7

2 Round 68.07 to the nearest whole number.

  a 68   **c** 69

  b 68.1   **d** 70

3 Round 234.34 to the nearest ten.

  a 230   **c** 234.3

  b 234   **d** 240

4 Round 1966.56 to the nearest ten.

  a 1966   **c** 1967

  b 1966.6   **d** 1970

5 Round 2962 to the nearest hundred.

  a 2900   **c** 2970

  b 2960   **d** 3000

6 Round 73 684 to the nearest hundred.

  a 73 680   **c** 73 700

  b 73 690   **d** 74 000

7 Round 24 763 to the nearest thousand.

  a 24 000   **c** 24 800

  b 24 760   **d** 25 000

8 Round 69.072 to the one decimal place.

  a 69.0   **c** 69.1

  b 69.07   **d** 70

9 Round 379.232 to one decimal place.

  a 379.0   **c** 379.23

  b 379.2   **d** 379.3

10 Round 258.3156 to two decimal places.

  a 258.3   **c** 258.316

  b 258.31   **d** 258.32

Unit 209
Ratio and proportion

Introduction

We use proportion in everyday life. Think of sharing a pizza with your friend – do you take one slice each and then another slice each? This is known as direct proportion: one slice for your friend and one slice for you. We can write this 1 : 1.

It would still be direct proportion if you had 2 slices every time your friend had one slice. We could write this 2 : 1.

Ratios are also used to compare two or more quantities in a variety of situations including baking cakes, mixing cement and diluting concentrated liquids such as cordial or weed killer. In each of these the proportion of one ingredient to the other is stated and happens every time.

Scale drawings and scale models use a stated ratio for the drawing. They are used in many different situations where it is important to plan or design something before making it.

- Plans of rooms help design the layout of the furniture.
- Plans or models of buildings show exact details of sizes of walls and windows.
- Diagrams or models of machines show their shape and size compared to other objects.

A map is a kind of scale drawing that shows where towns and places are located or where buildings and roads are located depending on the scale of the map.

Learning objectives

In this unit you will find information about:

- ratio and proportion.

This will help you to prepare for questions about:

- solving simple proportion problems
- working out quantities
- reading and using scales on maps and plans.

Solving simple proportion problems

Sometimes we don't know how much one item costs but we know how much a pack of four or six costs. We use ratio and proportion to work out the cost of one item. This assumes that each item costs the same.

If four items cost $4, then one of these items would cost 4 ÷ 4 = $1

Example 1

Four wooden boxes cost $200.
a How much does one box cost?
b How much would 8 boxes cost at the same rate?

Solution

a 200 ÷ 4 = $50 for one box.
b 8 × 50 = $400 for 8 boxes.

Example 2

Ten sheets of letter size paper cost $12.
a How much does one sheet cost?
b How much would six sheets cost at the same rate?

Solution

a 12 ÷ 10 = $1.20 for one sheet.
b 6 × 1.20 = $7.20 for 6 sheets.

Example 3

A hotel charges $240 for 3 nights.
a How much does the hotel charge for one night?
b How much would 10 nights at the hotel cost at the same rate?

Solution

a 240 ÷ 3 = $80 for one night.
b 10 × 80 = $800 for 10 nights.

Using a ratio

If you have six sweets and you share them equally between you and your friend, you would both have three sweets.

This could be written as a ratio 3 : 3, meaning 3 for one person and 3 for the other person.

As these are both the same number, we can divide each by 3 and write the ratio in a simpler form:

1 : 1

Ratios are usually written as simply as possible.

Example 1

If you always share your sweets in a ratio of $1:1$, we can work out how many sweets your friend will have if there are 10 sweets.

Solution

You : your friend

1 : 1 (2 sweets in total)

2 : 2 (4 sweets in total)

3 : 3 (6 sweets in total)

4 : 4 (8 sweets in total)

5 : 5 (10 sweets in total).

You will have 5 sweets and your friend will have 5 sweets.

You can work this out another way:

The number of sweets in each share is 2 ($1:1$, one for you and one for your friend), so you can divide the total number of sweets by the amount given each time they are shared.

$10 \div 2 = 5$

Example 2

A mother dilutes a juice drink concentrate with water for her children in the ratio of

1 unit of concentrate : 5 units of water

It does not matter what the unit of measure is, the ratio says that
1 spoonful of concentrate will be diluted with 5 spoonfuls of water,
1 cup of concentrate will be diluted with 5 cups of water and 1 litre of concentrate will be diluted with 5 litres of water.

If the mother has 3 units of concentrate, how much water will she need?

Solution

1 unit : 5 units

2 units : 10 units

3 units : 15 units

We can see that if she has 3 units of concentrate she will need 15 units of water to dilute it.

Or, the ratio is $1:5$ and the mother is using 3 units, so we can say

1 : 5

3 : ?

Keep the same ratio $1 \times 3 : 5 \times 3$

which gives us 3 units : 15 units.

It is important to note that $1:5$ is not the same as $5:1$.

$5:1$ would mean 5 units of concentrate to 1 unit of water.

Example 3

A man decides to give some money to two different charities in the ratio of 3 : 2

If he gives $30 to the first charity, how much money will he give to the second charity?

Solution

3 : 2

30 : ?

Keep the same ratio $3 \times 10 : 2 \times 10$

which gives us $30 for the first charity : $20 for the second charity.

Using a scale on maps and plans

It's possible to do a drawing of an object, town or country not to scale and, if it is clearly labelled with all the sizes, it may be useful. However, a scale drawing is much more useful because it is exactly the same shape as the real thing, but smaller. Making a scale drawing is like shrinking an object many times so that it can be shown on an ordinary piece of paper.

The sizes shown on the scale drawing will be in the same proportions as in real life. For example, a window which is half the length of a wall in real life, will be half the length of the wall on a scale drawing. This is important, as it means that the layout of the room on the scale drawing will work in real life.

You can also take measurements from a scale drawing and work out the actual measurements from them.

An example of a simple scale is 1 centimetre represents 1 metre.

This means that every centimetre on the drawing is 1 metre in real life.

A wall 8 centimetres long on the drawing is 8 metres long in real life.

1 centimetre represents 1 metre (100 cm) can also be written as a ratio 1 : 100.

This means that every unit on the drawing is 100 units in real life. We would probably think of this as 1 cm represents 1 metre but this could be any unit such as 1 mm represents 100 mm or 1 inch represents 100 inches.

There are many other scales depending on the size of the original object and the paper used to represent it.

2 centimetre represents 1 metre means that every 2 centimetres on the drawing is 1 metre in real life.

1 centimetre on the drawing is 0.5 metres in real life.

So, a wall 8 centimetres long on the drawing is 4 metres long in real life (8×0.5).

2 centimetre represents 1 metre can also be written as a ratio 1 : 50.

What would 4 centimetres represents 1 metre mean?

4 centimetre represents 1 metre means that every 4 centimetre on the drawing is 1 metre in real life.

1 centimetre on the drawing is 0.25 metres in real life.

Real world maths

Architects use scales when they are drafting designs for buildings. The same goes for surveyors when they draw diagrams of a piece of land surveyed for domestic or commercial use.

So, a wall 8 centimetres long on the drawing is 2 metres long in real life (8×0.25).

4 centimetre represents 1 metre can also be written as a ratio $1:25$

> ### Activity
>
> Look at the three scales mentioned above. Which scale would you use to draw a wall 15 m long on a piece of paper the size of this book?

The bigger the object in real life, the larger the number needed for the ratio. A map could have a scale of $1:100\,000$ which is 1 cm represents $100\,000$ cm (which is 1 km) so that the distances between places could be shown. Or the map could cover a smaller area of land and have a scale of $1:50\,000$ which would be 1 cm represents $50\,000$ cm or $\frac{1}{2}$ km.

Working out actual sizes from a scale drawing

To work out actual sizes, we start with

- the size on the scale drawing and
- the scale of the drawing.

Then either multiply or divide depending on the scale.

> ### Example 1
>
> A drawing of a room has a scale where 4 centimetres represent 1 metre.
>
> A wall of the room is 14 cm on the scale drawing.
>
> How long is the actual wall?
>
> ### Solution
>
> The scale says that every 4 cm on the drawing is 1 m in real life.
>
> We need to know how many lots of 4 cm there are in the length of the wall on the drawing – that will be the number of metres in real life.
>
> To find how many lots of 4 there are in 14, we could subtract 4s:
>
> $14 - 4 = 10$ (that's one lot of four)
>
> $10 - 4 = 6$ (that's 2 lots of four)
>
> $6 - 4 = 2$ (that's 3 lots of four)
>
> 2 is half of four (that's $3\frac{1}{2}$ lots of four)
>
> So, the actual wall is 3.5 metres long.
>
> We could also do this calculation by dividing:
>
> $14 \div 4 = 3.5$
>
> Here is another way of thinking about the scale:
>
> Where 4 cm represent 1 metre, 1 cm represents $\frac{1}{4}$ of a metre.
>
> So we need to find a quarter of the length on the scale drawing to work out the actual length.
>
> A quarter of 14 is the same as $14 \div 4 = 3.5$

Usually, the numbers used in scales for maps or plans are even numbers or multiples of 5 or 10 as these numbers are simpler to work with.

Look at Unit 4, Operations on decimal fractions, if you need to refresh your memory about division.

Example 2

A drawing of a building plot has a scale where 1 centimetre represents 5 metres.

One side of the plot is 3.5 centimetres on the scale drawing.

What is the actual length of this side?

Solution

The scale says that every 1 cm on the drawing is 5 m in real life.

The number of centimetres on the drawing is the number of lots of 5 metres there are in the actual length.

To find 3.5 lots of 5 metres, add three fives and then half of five:

$5 + 5 + 5 + 2.5 = 17.5$

The side of the plot is 17.5 metres.

Or, multiply 3.5 by 5:

$3.5 \times 5 = 17.5$

Example 3

The scale on the plan is $1 : 50$.

A wall measures 3 m in real life. How big should this be drawn on the plan?

Solution

$1 : 50$ means 1 unit represents 50 units. We can say 1 cm represents 50 cm (or $\frac{1}{2}$ m).

The measurement of the wall is in metres but the measurement on the plan is likely to be centimetres. Remember there are 100 cm in a metre.

If 1 cm represents 50 cm ($\frac{1}{2}$ m)

Then 2 cm represents 100 cm (or 1 m)

3 cm represents 150 cm (or $1\frac{1}{2}$ m)

4 cm represents 200 cm (or 2 m)

5 cm represents 250 cm (or $2\frac{1}{2}$ m)

6 cm represents 300 cm (or 3 m)

Or, we could divide 3 m by 50 cm – make sure the units are the same before you divide.

$300 \div 50 = 6$

Tasks

1 A meal cost $100 for 4 people.

   a How much does the meal cost for one person?

   b How much would a meal for 6 people cost at the same rate?

2 A cleaner takes 30 minutes to clean 3 bedrooms.

   a How long does it take the cleaner to clean one bedroom?

   b How long would it take the cleaner to clean ten bedrooms at the same rate?

   (Remember there are 60 minutes in 1 hour.)

3 Glenmore and Khenan share a bag of sweets in the ratio 1 : 4.
   There are 20 sweets.

   a How many sweets does Glenmore have?

   b How many sweets does Khenan have?

4 A hotel charges $600 for 5 nights.

   a How much does the hotel charge for one night?

   b How much would 3 nights at the hotel cost at the same rate?

5 Abigay draws a plan of her bathroom floor.
   The bathroom floor is a rectangle 3 m by 4 m.
   Abigay uses a scale of 1 : 20.
   What will the size of the rectangle be on the plan?

6 A drawing has a scale where 1 centimetre represents 5 metres.
   One side of the plot is 5 centimetres on the scale drawing. What is the actual length of this side?

7 A drawing has a scale of 1 : 100.
   A wall is shown as 8 cm on the plan.
   What is the actual length of the wall?

8 A window measures 100 cm × 80 cm.
   A man draws a plan with a scale of 1 : 20
   What will the size of the window be on the plan?

9 A street map has a scale where 1 centimetre represents 50 metres. On the map, the distance between the station and a hotel is 5 cm.
   What is the actual distance?

10 A designer makes a scale drawing of a new sports arena.
    The drawing has a scale where 1 centimetre represents 4 metres.
    The pitch is 106 m long.
    What is the length of the pitch on the scale drawing?

Test your knowledge

1 A plan of a kitchen has a scale where 4 cm represent 1 m.
   One wall has a length of 12 cm on the plan.
   What is the actual length of the wall?

   a 3 m

   b 7 m

   c 8 m

   d 12 m

2 A drawing of a park has a scale where 1 cm represents 2 m.
   The actual park has a length of 24 m.
   What is the length of the park on the drawing?

   a 12 cm

   b 24 cm

   c 36 cm

   d 48 cm

3 A model of a building has a scale where 1 cm represents 5 m.
   The actual building has a height of 120 m.
   What is the height of the scale model of the building?

   a 20 cm

   b 24 cm

   c 60 cm

   d 72 cm

4 A man uses a map with a scale of 1 cm to 5 km.
   On the map, the distance is 15 cm.
   What is the actual length of the journey?

   a 3 km

   b 7.5 km

   c 30 km

   d 75 km

5 A cook uses 1 egg to make 12 small cakes.
   How many cakes will 6 eggs make?

   a 2

   b 6

   c 12

   d 72

6 A hairdresser mixes colour to developer in the ratio of 1 : 2 to dye hair.
She uses 100 ml of colour.
How much developer should she use?

a 50 ml

b 100 ml

c 200 ml

d 300 ml

7 Grace and Levi share a bag of sweets in the ratio 2 : 3.
There are 20 sweets.
How many sweets does Grace have?

a 4

b 5

c 8

d 10

8 A hotel charges $240 for 2 nights.
How much would 6 nights at the hotel cost at the same rate?

a $120

b $360

c $720

d $1440

9 Six cards cost $12.
How much would four cards cost at the same rate?

a $2

b $3

c $8

d $18

10 A reader has 20 fiction books and 10 non-fiction books.
What is the ratio of fiction books to non-fiction books?

a 1 : 2

b 2 : 1

c 1 : 3

d 3 : 1

Introduction

Understanding what the word 'average' means and how to calculate an average is a very important skill in mathematics. If you look up 'average' in the dictionary, you can see it means typical, ordinary, medium and middle. In mathematics, we have different ways of calculating an average.

Averages are used to make all sorts of decisions that are important to our day-to-day lives at home and at work.

We need to be accurate when calculating averages: one mistake with your calculator could result in working out the average height of a male as 6 ft 8 inches (203 cm) or the average shoe size being size 15.

Such results might be true for international basketball players, but would lead to an interesting meeting with your boss if you used them to place the company's next order of safety clothing! These examples highlight the importance of checking whether your answers make sense when you have worked out your calculations.

As well as finding the average, it is often useful to find the 'range' of the numbers. Range means the full extent covered.

You will see the term 'range' used in lots of different places. Shops have a large range of sizes, colours, makes and models which gives us lots of choice. Leisure venues and holiday parks advertise a wide 'range' of family activities.

The terms 'ranging from' is also often used in shops. A furniture store, for example, may advertise prices ranging from $199 to $2599 so that potential customers know that they stock products suiting all incomes. Garages might advertise a range of cheap and expensive car brands to let customers know that they will be able to find a car they can afford to buy.

Within many workplaces, range provides a description which helps us to plan in advance. Between 200 and 250 customers each lunchtime in a restaurant, 35 to 40 cars repaired in a garage and 2 to 4 drink breaks could be approximate numbers used to plan finances, make orders and organise staffing in different workplaces.

Real world maths

Shops, for example, work out the average number of customers and use this information to make sure they have enough staff working at different times and to order the right amount of stock. In car and cycle workshops, the average time taken to complete a service plays an important part in setting pricing. Averages impact on us in lots of ways at home too, for example informing costs for insurance policies.

Learning objectives

In this unit you will find information on:

- average and range.

This will help you to prepare for questions about:

- how to calculate average and find range.

Mean

When asked to work out a mathematical average, most people would probably use the following calculation:

'Add up all the number values and then divide this total by how many numbers you have.'

This is the most commonly used average. It is the only average required at Stage 2 and is called the **arithmetic mean average**. There are actually three averages and the other two, along with the reasons they are useful, are included at Stage 3.

> At this level, you will usually only be asked to calculate the mean of up to ten numbers.

The mean average of a set of numbers can be worked out as follows:

$$\text{mean average} = \frac{\text{the sum of all the numbers}}{\text{how many numbers there are}}$$

Let us look at an example of using this method to work out the mean average.

Find the mean of 6, 8, 10, 4, 2

Step 1: add up the total of the numbers $(6 + 8 + 10 + 4 + 2 = 30)$

Step 2: divide the total (30) by how many numbers there are (5) = $30 \div 5 = 6$

So, the mean average is 6.

> You do not need to order the numbers when calculating the mean. The total of $6 + 8 + 10 + 4 + 2$ is exactly the same as the total of $2 + 4 + 6 + 8 + 10$.

Of course, you will not always just be given some numbers and asked to find the mean. The question will often be set in a context, for example the numbers given (6, 8, 10, 4, 2) could be the number of hours overtime worked in a five-week period. The context does not change the method you need to use in any way; the only difference would be that the answer should be given as 6 hours.

> **Tip for assessment**
>
> The answer to the mean calculation is always a number more than the lowest number and less than the highest number.

> **Tip for assessment**
>
> A common error made when calculating the mean is to forget to use the 0 if one is included in the numbers you are given.
>
> If, for example, the number of hours of overtime over four weeks were 8 hours, 0 hours, 7 hours and 5 hours then divide the total (8 + 0 + 7 + 5 = 20) by 4 to find the answer of 5 hours.
>
> If you divide by 3, you are finding the average overtime over 3 weeks.

In an exam, you may be given the total and asked to calculate the mean. For example:

Alan worked 20 hours overtime over 4 weeks. What was the mean average number of hours of overtime worked each week?

In this case, as step 1 has already been calculated for you, you need only complete step 2 (division) to find the answer (20 hours $\div$ 4 weeks) = 5.

> **Example**
>
> Find the mean of 3, 2, 5, 4, 1, 3
>
> ## Solution
>
> Step 1: add up the total of the numbers
>
> $3 + 2 + 5 + 4 + 1 + 3 = 18$
>
> Step 2: divide the total (18) by how many numbers there are (6)
>
> $18 \div 6 = 3$
>
> So, the mean average is 3.

Range

The **range** is the difference between the highest value and the lowest value in a set of numbers.

The range of a group of numbers tells us how far the values are spread. It does not tell you anything about the size of the numbers, but does tell you how far apart they are. The range of ages in a town-centre night club, for example, is likely to be small compared to the age range at a football match as football matches are watched by people of all ages.

In mathematics, the range is calculated by subtracting the smallest number from the largest number. For example, if the oldest person in a nightclub is 28 years old and the youngest is 18 then the range would be:

$28 - 18 = 10$ years

If the eldest supporter at the football match was 87 years old and the youngest was 3, then the range would be:

$87 - 3 = 84$ years

Comparing the two ranges, you can say that the first set has ages which are quite close together, but in the second set they are very spread out.

> **Example**
>
> Find the range of the following car prices:
>
> $13 000, $15 500, $16 000, $12 000, $16 500
>
> ## Solution
>
> Step 1: find the highest price ($16 500)
>
> Step 2: find the lowest price ($12 000)
>
> Step 3: subtract the lowest from the highest ($16 500 − $12 000 = $4 500)
>
> The range is $4 500.

> **Tip for assessment**
>
> A common error is to write the range as between 16 500 − 12 000 or $16 500 − $12 000. Remember that you need to work out the difference between the highest and lowest values.

Tasks

1 Find the mean and the range of each of these sets of data.

   a 1, 2, 3, 6

   b 1 m, 3 m, 5 m, 7 m, 9 m

   c 12 cm, 4 cm, 6 cm, 8 cm, 5 cm, 13 cm

2 Find the range of the following numbers.
15, 82, 41, 77, 89, 6, 56

3 Calculate the mean of the following numbers.
17, 27, 11, 29, 36

4 Here are the times, in minutes, for a bus journey between two towns.
15, 7, 9, 12, 9, 19, 16, 11, 9, 18

   a What is the mean time for the journey?

   b Find the range of these times.

5 The ages of the employees in a local business are as follows:
46, 52, 38, 25, 31, 62
What is the range of the ages?

6 Twelve people have their hand span measured so that their employer can provide safety gloves. The measurements are shown below in millimetres:
225, 216, 188, 212, 205, 198, 194, 180, 194, 198, 200, 194

   a What is the mean hand span?

   b What is the range of the hand spans?

   c Comment on what this will mean for the company when it orders the gloves.

7 Here are the results of a survey where parents and children wrote down how many hours a week they spent watching television:

Children	15	14	20	17	22	21	23	16	12	15
Parents	9	13	15	18	0	10	12			

   a Find the mean and range of these times.

   b On average, do the parents or children watch more television?

8 A gardener measures the height in centimetres of her sunflower plants. These are the heights:
140, 123, 131, 89, 125, 123, 115, 138

   a What is the mean height?   **b** What is the range of the heights?

9 The salaries of ten workers are as follows:
$20 000, $20 000, $20 000, $18 000, $22 000, $23 000, $25 000, $21 000, $23 000, $28 000

   a What is the range of salaries?

   b What is the mean salary?

   c The most highly-paid employee earns $28 000. If she has a pay rise of $5 000, how does the mean salary change?

10 Tom and Freya go bowling. These are their scores:

Tom	7	8	5	3	7
Freya	10	8	3	1	3

   a Find the mean and range of each person's scores.

   b Write down two comments about their scores.

Test your knowledge

1   The table shows the forecast for the temperature each day during a weekend break in Europe.

Day	Temperature (°C)
Friday	20
Saturday	24
Sunday	25
Monday	23

What is the average (mean) temperature in °C?

a 24.5   **c** 23

b 24   **d** 20

2   What is the range of temperatures over the same weekend?

Day	Temperature (°C)
Friday	20
Saturday	24
Sunday	25
Monday	23

a 3 °C

b 4 °C

c 5 °C

d 20 °C

3   A company sales representative visits 160 customers in one month.
The 160 journeys cover 3200 miles altogether.
What is the average (mean) journey distance?

a 2 miles

b 20 miles

c 0.2 miles

d 512 miles

4   A leading football club records the size of the crowd during the first six matches of the season. This information is shown in the table below.

Match	1	2	3	4	5	6
Crowd size in 100s	44	42	39	44	37	40

What is the average (mean) crowd size over those six games?

a 40 000

b 41 000

c 42 000

d 44 000

5 A call-centre operator answers one call as soon as she finishes the previous call.

She answers 15 calls in 1 hour.

What is the average (mean) time taken for each call?

a 15 minutes

b 5 minutes

c 4 minutes

d 3 minutes

6 A dressmaker has five lengths of ribbon.

They measure 2 m, 4 m, 1 m, 4 m and 6 m.

What is the range of lengths?

a 2 m

b 4 m

c 5 m

d 6 m

7 Find the average (mean) of these weights:

3 kg, 7 kg, 10 kg, 8 kg, 3 kg, 11 kg

a 3 kg

b 7 kg

c 8 kg

d 9 kg

8 An item is on sale in four different shops at different prices:

$15 $27 $30 $42

What is the range of prices?

a $12

b $13

c $26

d $27

9 A group of 10 students sit a test.

Here are their scores: 6, 7, 7, 9, 5, 7, 7, 7, 10, 8

What is the range of the scores?

a 2

b 5

c 7

d 10

10 A part-time worker records the time (in hours) he works over 5 days.

Monday	Tuesday	Wednesday	Thursday	Friday
5	4	0	6	5

What is the average (mean) time he works each day?

a 3 hours

c 5 hours

b 4 hours

d 6 hours

Introduction

People sometimes think algebra is complicated but you have already substituted values into an equation. An equation simply says that the two things either side of an equals sign (=) are equal.

Do you remember finding the area of this shape in Unit 104?

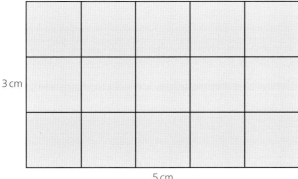

3 cm

5 cm

In Unit 104 we found the area of a rectangle initially by counting up the squares and also by multiplying $5 \times 3 = 15\,\text{cm}^2$.

The formula for finding area was length × width, and you multiplied the two measurements together to find the area. In other words, you substituted the value of the length instead of saying 'length' and the value of the width instead of saying 'width'.

In Unit 101, we looked at pairs of numbers that add up to 10. A typical question was 6 + ? = 10. If you can solve this and give the answer 4, you are solving a simple equation with one unknown (the ?).

Learning objectives

In this unit you will find information on

- solving simple equations.

This will prepare you for questions about:

- how to substitute values into an equation expressed in words or simple formulae and solving simple equations with one unknown.

This unit builds on your work in previous units, so don't be frightened by the title of the unit.

Substituting values into an equation

Example 1

The selling price of an item is the total of the cost of production and the profit, so the formula is

selling price = production cost + profit

What is the selling price if the production costs are $4 and the profit is $2?

Solution

First substitute the values into the formula:

selling price = 4 + 2

so selling price = $6

Formulae can be written in words but they are also written using letters or symbols to represent the words.

The equation for finding the area of a rectangle is always **length × width** but sometimes this is shortened to area = $\ell \times w$.

Example 2

Find the area of a room where $\ell = 4\,\text{m}$ and $w = 3\,\text{m}$.

Area = $\ell \times w$

Solution

First substitute the values into the formula:

area = 4 × 3

so area = $12\,\text{m}^2$

Example 3

Find the perimeter of a yard where $\ell = 5\,\text{m}$ and $w = 4\,\text{m}$.

Perimeter = $2\ell + 2w$

Solution

First substitute the values into the formula:

perimeter = (2 × 5) + (2 × 4)

so perimeter = 10 + 8

perimeter = $18\,\text{m}$

Solving simple equations with one unknown

In Unit 101, we used different symbols to stand for an unknown number, such as $3 + ? = 9$.

If $3 + ? = 9$, we can probably guess that $? = 6$.

We could work this out as an equation.

Example 1

$3 + ? = 9$. What is the value of $?$?

Solution

To solve this equation you need to have $?$ by itself on one side of the $=$ sign.

If you take 3 away from $3 + ?$ this leaves $?$ by itself.

Then you need to do the same to the other side of the $=$ sign.

$9 - 3$

Now you have

$? = 9 - 3$

so $? = 6$

We can check this by putting 6 as the value of $?$ in the original equation

$3 + 6 = 9$ This is correct so we know the answer $? = 6$ is correct.

In Unit 101, we used $?$ and y to represent a missing number. Other letters and symbols can be used, for example x is often used to represent a missing number.

Example 2

$x - 1 = 5$

What is the value of x?

Solution

To solve this equation you need to have x by itself on one side of the $=$ sign.

If you add 1 to $x - 1$ this leaves x by itself.

Then you need to do the same to the other side of the $=$ sign: $5 + 1$

Now you have

$x = 5 + 1$, so $x = 6$

We can check this by putting 6 as the value of x in the original equation: $6 - 1 = 5$.

This is correct so we know the answer $x = 6$ is correct.

Example 3

$3a - 2 = 25$

What is the value of a?

Solution

To solve this equation you need to have a by itself on one side of the = sign.

If you add 2 to $3a - 2$ this leaves $3a$ by itself.

Then you need to do the same to the other side of the = sign:

$25 + 2$

$3a = 27$

$3a$ is $3 \times a$ and the opposite of multiply is divide so we divide each side by 3

$\dfrac{3a}{3} = \dfrac{27}{3}$

$a = 9$

We can check this by putting 9 as the value of a in the original equation:

$3a - 2 = 25$

$(3 \times 9) - 2 = 25$. This is correct so we know the answer $a = 9$ is correct.

Tasks

1 Find the area of a rectangle where $\ell = 6\,\text{cm}$ and $w = 8\,\text{cm}$.
   Area $= \ell \times w$

2 The charge to hire a hall is $10 plus a charge of $7 per hour.
   How much will it cost to hire the hall for 3 hours?

3 What is the selling price if production costs are $24 and profit is $10?
   selling price = production cost + profit

4 Solve $p - 1 = 7$

5 Solve $2y + 3 = 11$

6 Solve $3x - 2 = 25$

7 Solve $6 + w = 10$

8 Find the distance travelled at 50 kilometres per hour, if you travel for 2 hours.
   Distance = time $\times$ speed

9 Solve $5a + 2 = 37$

10 Solve $10 - b = 4$

Test your knowledge

1 Calculate the profit of a business using this formula:
profit = revenue − costs
Revenue is $14 600 and costs are $9 500.

   a $5 100   **c** $14 600

   b $9 500   **d** $24 100

2 Find the area of a yard where $\ell = 20$ m and $w = 15$ m.
Area $= \ell \times w$

   a 35 m^2 **c** 210 m^2

   b 70 m^2 **d** 300 m^2

3 $2y = 6$
What is the value of y?

   a 2   **c** 6

   b 3   **d** 8

4 Solve $p − 5 = 2$.

   a 2   **c** 5

   b 3   **d** 7

5 Solve $2x + 1 = 17$

   a 8   **c** 14

   b 9   **d** 16

6 Solve $4y − 4 = 40$

   a 4   **c** 10

   b 9   **d** 11

7 Solve $6 + 2w = 10$

   a 2   **c** 4

   b 3   **d** 5

8 Find the distance travelled at 60 kilometres per hour, if you travel
for 3 hours.
Distance = time × speed

   a 20 km   **c** 120 km

   b 30 km   **d** 180 km

9 Solve $5a + 10 = 40$

   a 5   **c** 8

   b 6   **d** 10

10 Solve $12 − b = 3$

   a 4   **c** 12

   b 9   **d** 15

Unit 212
Shape and space

Introduction

In Unit 104 we looked at some 2D shapes such as squares, rectangles and triangles and tessellation of these shapes. We also looked at the nets of simple cuboids. This unit builds on these ideas and looks at measuring angles and finding the size of different angles by using the properties of shapes and angles on a straight line.

We also used the idea of symmetry and reflection and this unit takes this concept further by looking at transformations.

Unit 104 introduced area and perimeter of simple shapes like rectangles and this unit will build on the basic shapes to find the area and perimeter of L-shaped rooms. We also find the volume of cuboids.

You may find it useful to look through Unit 104 before you start working on this unit.

Learning objectives

In this unit you will find more information on:

- measuring angles
- calculating perimeter, area and volume of specific shapes.

This will help you to prepare for questions about:

- how to work with shape and space to find angles and calculate perimeter, area and volume.

Measuring and drawing shapes

You will be familiar with these shapes from Unit 104. At Stage 2, you should be able to recognise and draw these shapes.

square

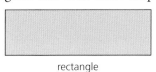

rectangle

cube

sphere

circle

cuboid

This 3D shape is a cylinder.

Measuring angles

You know that a square and a rectangle have four right-angles or corners.

A right-angle measures 90°.

There are 360° in a circle and 360° in a square (4 corners each 90°).

We use a protractor to measure angles. This image shows a protractor with the red line indicating 90°.

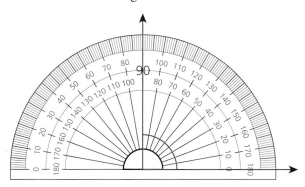

This protractor only shows 180° because it is only half of the circle. Some protractors are circles and show 360°.

Example 1

What is this angle?

Solution

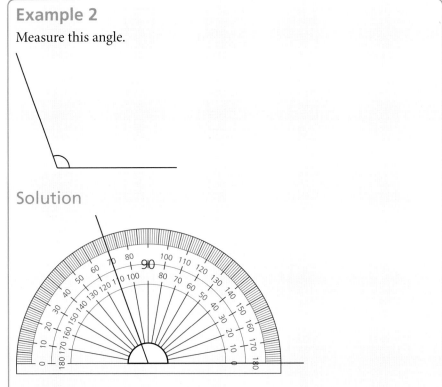

We can place the protractor on top of the angle and read off from the inside measure. The reading is exactly halfway between 40 and 50, so the angle is 45°.

Example 2

Measure this angle.

Solution

We can place the protractor on top of the angle and read off from the inside measure. The reading is 110°.

Finding missing angles

We can also work out angles from our knowledge of shapes.

If the triangle has three equal angles, what size would they each be? ←

180 divided by 3 = 60°

An **equilateral triangle** has three equal sides and three equal angles.

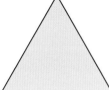

Each angle is 60°.

Not all triangles are equilateral.

> The angles of a four-sided shape, e.g. a square, always add up to 360°.
>
> The angles of a three-sided shape, i.e. a triangle, always add up to 180°.

Activity

Measure the angles of these triangles.

Do they add up to 180°?

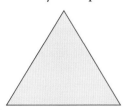 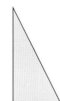

This triangle has two sides the same but the bottom is a different length.

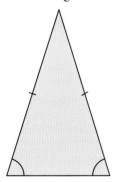

An **isosceles triangle** has at least two equal sides and at least two equal angles.

This is an isosceles triangle. In this triangle two sides are equal and two angles are equal.

An equilateral triangle is a special isosceles triangle.

Example 1

Find the missing angle *A* in this triangle.

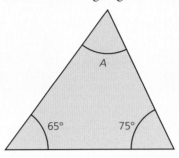

Solution

The angles of a triangle always add up to 180°.

65 + 75 + *A* = 180 or we can say *A* = 180 − 65 − 75

so *A* = 40°

Example 2

Find the missing angle *x* in this triangle.

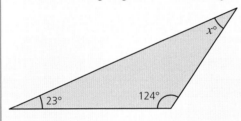

Solution

The angles of a triangle always add up to 180°.

23 + 124 + *x* = 180 or we can say *x* = 180 − 23 − 124

x = 33°

We can also use the properties of angles on a straight line to find the size of an angle.

Example 3

Find angle *a*.

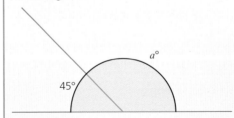

Solution

Angles on a straight line add to 180°.

180 − 45 = 135°

Example 4

Find angle p.

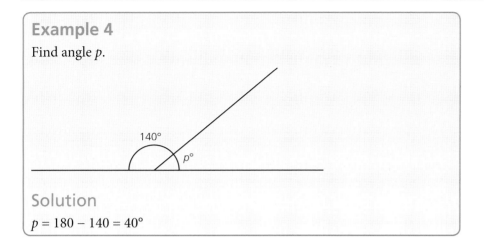

Solution

$p = 180 - 140 = 40°$

Transformations

Shapes can be transformed in different ways such as

- **translation** – moving a shape up, down or across but not altering it in any way
- **rotation** – turning the shape round
- **reflection** – like a reflection in a mirror
- **enlargement** – this is described by its scale factor and is covered in Stage 3.

At Stage 2 you need to recognise **congruent** shapes. Congruent means identical in shape and size so congruent does not include shapes that have been enlarged or reduced in size.

Example 1

A   B   C

D   E   F

a Which of these shapes are examples of translation?
b Which of these shapes are examples of reflection?
c Which of these shapes are examples of rotation?
d Which of these shapes are congruent?

Solution

a A and D. The shape is exactly the same. It is just in a different place.
b D and E or A and E
c A, B and C or D, B and C
d A, B, C, D, E
F is not congruent because it is a different size.

Perimeter

When talking about length and width, people do not agree. Some people think that length should always be a horizontal measurement and width should be a vertical one. The more common view is that the length should always be the longest measurement and the width should be the shortest.

At Stage 1 we said that the **perimeter** is the length of measurement around an entire two-dimensional shape. Think of the perimeter as the surround, or boundary, of these shapes where the total measurements of each width and length are added together. One way to remember what the perimeter of a shape means is to think of:

p for path – a path around the shape.

When working out the perimeter, it is important to use a suitable unit of measurement. For example, for the perimeter of a matchbox you could use centimetres and millimetres, for a bedroom you could use metres and centimetres and for a large area of land you would use kilometres.

Let's try some worked examples.

Example 1

Calculate the perimeter of the following room.

3 m

5 m

Solution

Add the lengths of each side:

$5 + 3 + 5 + 3 = 16\,m$

So, the perimeter of the room = 16 metres

Example 2

Calculate the perimeter of the following field.

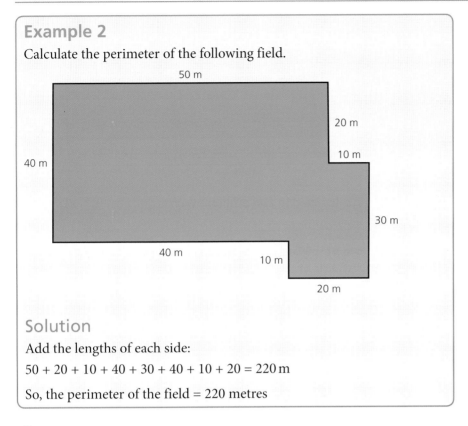

Solution

Add the lengths of each side:

50 + 20 + 10 + 40 + 30 + 40 + 10 + 20 = 220 m

So, the perimeter of the field = 220 metres

Tip for assessment

Remember to identify the unit of measurement (in this instance m for metre) in your final answer.

Exam questions often omit some measurements on purpose. You must calculate these using the other measurements available and include them in your calculation.

Learner tip

If calculating for a particular material, such as fencing, remember not to enter any section of the perimeter marked as walls or doors into the sum.

Area

The area of a 2D shape is the space within a flat surface. This is calculated by multiplying the length and width together on a standard shape such as a rectangle.

For a rectangular room measuring 5 m (length) by 3 m (width), the area would therefore be $5 \times 3 = 15 \, m^2$.

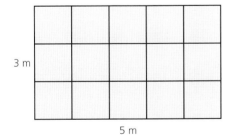

Remember the small '2' represents squared, indicating that there is an overall area consisting of 15 squares, with each square measuring 1 metre on all four sides. This is easier to understand by looking at this diagram.

At Stage 2, calculating area is normally related to **composite shapes**. Composite shapes, or compounds, are more complex shapes that require breaking down into more than one regular shape, such as rectangles and squares.

Example 1

Find the area of this shape.

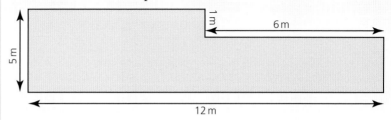

The area can be calculated in two different ways

Method A

Calculate the area of the large rectangle and take away the shaded area.

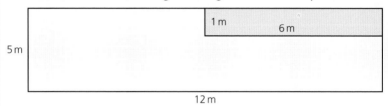

The large rectangle is $12\,m \times 5\,m = 60\,m^2$

The shaded area is $6\,m \times 1\,m = 6\,m^2$

Area $= 60 - 6 = 54\,m^2$

Method B

Split the shape into two rectangles A and B.

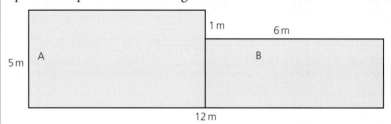

Rectangle A. We know one measurement 5 m but we don't know the other measurement.

The total length is 12 m and the top of rectangle B is 6 m so the length of A is $12 - 6$.

Area of A $= 5 \times 6 = 30\,m^2$

Area of B $= 4 \times 6 = 24\,m^2$

Total area $=$ A $+$ B $= 30 + 24 = 54\,m^2$

Use whichever method you find easier in your own calculations.

Example 2

Find the area of the following composite shape.

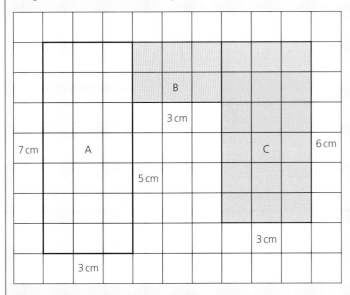

Solution

Step 1: Divide the composite shape into regular shapes.

Shape A = 7 cm × 3 cm, Shape B = 3 cm × 2 cm, Shape C = 6 cm × 3 cm.

Step 2: Calculate the area of each regular shape created.

A = 7 × 3 = 21

B = 3 × 2 = 6

C = 6 × 3 = 18

Step 3: Add each of three shape areas.

A + B + C = 21 + 6 + 18 = 45

Step 4: Check the units; they are all centimetres so it is cm².

The area of the composite shape is 45 cm².

Learner tip

When calculating the area of a composite shape, it can sometimes be easier to work out the area of two or more small shapes, or work out the full area of a shape and then subtract from this the area not included. Either method is acceptable.

Calculating the volume of 3D shapes

If we say a shape is 3D we mean that it has three dimensions: length, width and depth.

We calculated the volume of small cubes and cuboids in Unit 104.

Volume of a cube or cuboid = **length × width × height** (that is, area of flat surface × height)

Look at this cuboid.

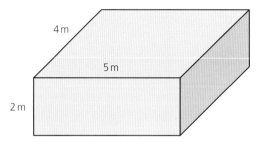

Therefore, the volume is $5\,m \times 4\,m \times 2\,m = 40\,m^3$. Remember for area we multiplied two measurements (length and width) and the answer was in squared units such as m^2.

For volume we have multiplied three measurements (length, width and height) and the small '3' represents cubic metres. This shows that there is an overall volume of 40 square cubes, with each cube measuring 1 m in length, width and height.

> **Tip for assessment**
>
> Always remember to indicate whether an answer is squared (^{2}) for area or cubed (^{3}) for volume.
>
> The terms **volume** and **capacity** are commonly confused. Volume is the space taken by a 3D object, whereas capacity refers to the available space within a container.

Tasks

1 What does this angle measure?

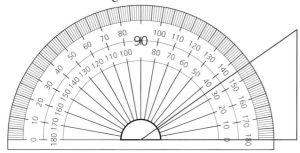

2 What is the value of angle *y*?

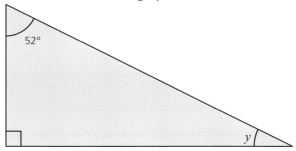

52°

y

3 What is the value of the angles *a*, *b* and *c* in the image below?

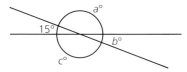

4 What is the difference between an isosceles triangle and an equilateral triangle?

5 Which of these shapes are congruent?

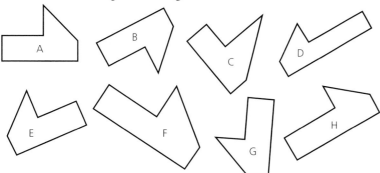

6 Calculate the area of this shape.

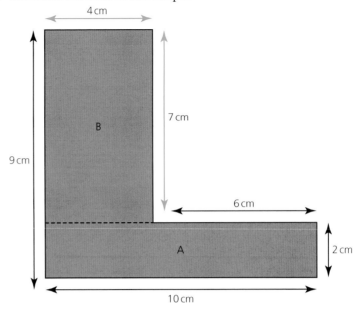

7 You need to buy a new carpet for an L-shaped room as shown below.

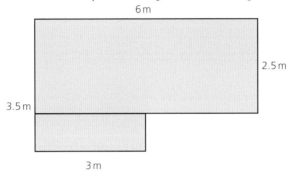

a What is the total area of the room?

b What is the perimeter of the room?

8 This is a shoe box.

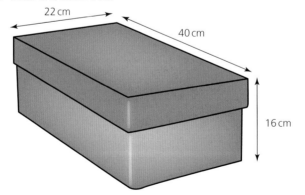

a What is the volume of the shoe box?

b What is the surface area of the bottom of the box?

Test your knowledge

1   What is the value of angle *h*?

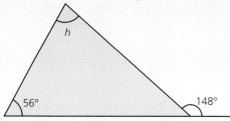

   a 24°

   b 56°

   c 92°

   d 156°

2   What is the value of angle *a*?

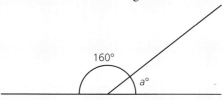

   a 20°

   b 40°

   c 140°

   d 360°

3   A fish tank has sides 15 cm × 60 cm × 25 cm long. What is its overall volume?

   a 552 500 cm³

   b 22 500 cm³

   c 925 cm³

   d 100 cm³

4   What is the perimeter of a rug which measures 2 m × 1.5 m?

   a 3 m

   b 3.5 m

   c 7 m

   d 8 m

5   What is the perimeter of a poster which measures 60 cm × 50 cm?

   a 110 cm

   b 220 cm

   c 300 cm

   d 3000 cm

6 You need to buy a new carpet for the room shown below. What is the total area of the room?

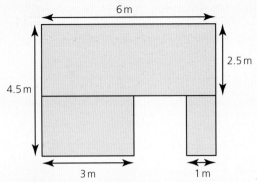

a 21 m²

b 23 m²

c 25 m²

d 27 m²

7 What is the perimeter of the room shown in question 6?

a 21 m

b 23 m

c 25 m

d 27 m

8 Look at the plan of a room below.
What is the area of the room?

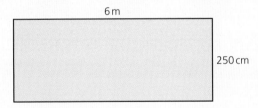

a 7.5 m²

b 12 m²

c 15 m²

d 17 m²

9 What is the perimeter of the room shown in question 8?

a 7.5 m

b 12 m

c 15 m

d 17 m

10 Which statement is not true?

a An isosceles triangle is an equilateral triangle.

b An equilateral triangle has three equal angles.

c An isosceles triangle has two equal sides.

d An equilateral triangle has angles of 60°.

Unit 213
Construct, interpret and use tables of figures, graphs, charts and maps

Introduction

This unit is about being able to read and understand the information that we meet every day.

At Stage 1, in Unit 103, we looked at four different ways of presenting information:

- pictograms
- tables (including lists)
- bar charts
- simple graphs.

This unit build on these skills and looks at more complex presentations of pictograms, tables, bar charts and line graphs and introduces:

- frequency tables
- pie charts
- maps.

They all present a summary of information in a visual way, which should be easy to understand!

At Stage 2, you will need to read and understand information in these different presentations and also be able to construct your own presentations to show information in different ways.

You may find it useful to look through Unit 103 before you start working on this unit.

Learning objectives

In this unit you will find information on how to:

- construct and interpret frequency tables, bar charts, pictograms, simple graphs and pie charts
- interpret information on tables
- read maps.

Tables

At Stage 2 you will extract information from tables similar to this one that we looked at in Stage 1.

Snack item	Weight (g)	Price ($)
Vanilla cookies	480	4.39
Wholewheat crackers	284	1.98
Butter cookies	55	0.55
Nacho cheese crackers	50	0.60
Sugar-free oatmeal cookies	227	4.40

> You may have tables with more columns or rows but the principle is the same.

Which snack is the most expensive?

The top row shows you what is in the table. You know there will be a list of snack items, their weight will be shown in grams (g) and the price in dollars ($). There is no need to show g after each weight or $ after each price as this is in the heading row.

The vanilla cookies have the highest price but they don't look so expensive now we can see the weight of them. We can see there are 480 g for $4.39, so by weight they are the cheapest snack (less than 1 cent per gram). Butter cookies look like the cheapest as the pack is the cheapest but there are 55 g for $0.55 (1 g for 1 cent).

Sugar-free oatmeal cookies are the most expensive. The pack is only slightly more expensive than vanilla cookies but you get less than half the weight.

Frequency tables

At Stage 1 we used a tally chart to record discrete data and counted the number of people who gave different colour choices. If we have a large number of items we may need to group the data and use a suitable class interval for each group.

Frequency tables are an efficient method of collecting and displaying total numbers of items or how often something occurs. The title would tell you what the results were from.

Mark	Tally	Frequency
0–5	||	2
6–10	||||	4
11–15	卌 |	6
16–20	卌 |||	8

This example might show one group's results for a mathematics test. The more you understand this table, the more likely you are to be one of the eight individuals scoring a mark in the 16–20 category!

There are three columns in this table. From left to right, the first shows the mark boundary, the second shows the tally and the final column shows the total frequency. The mark boundary is in whole numbers because no half marks were given on the test. It is important to start the next mark boundary with the next mark possible. If we had boundaries of 0–5 and 5–10, you would not know in which of the two groups to put someone who gained 5 marks.

> If you presented this information to an audience, they would not need to see the tally column as this is for collection. You would probably choose to display the information as a chart or graph.

Timetables

Managing your time is a very important skill, so it is common to see data handling questions which involve time. It is usually possible to use a number of different strategies to solve the tasks in exams, including trial and error, and there will often be more than one correct solution.

When planning rotas and schedules, it is a good idea to start by creating a table with the correct number of rows for input. The table below is suitable to input work schedule information for six employers for each day of the week. Include more or fewer rows depending on the number of employees and leave out weekends if they are not relevant. Remember to include a label for each column as in this example.

Employee work schedule

Week ending: _____

Employee	Hours						
	SUN	MON	TUES	WED	THUR	FRI	SAT

A similar type of table can be used to plan daily staffing rotas where you are given constraints about the number of staff that need to be working at a particular time. This helps to take into consideration lunch breaks and other routine breaks.

Bar charts

Bar charts present information as a series of blocks or bars. The height of each bar represents the amount for that particular category.

If the colours show something important, there will be a **key** to show you what each colour means. (The key is sometimes called a **legend**.)

The lines with numbers or names going across and up the page are called the **axes**. The line going across the page from left to right is the **horizontal axis** (imagine the horizon lying across the sea) and the one going up is called the **vertical axis**.

A standard bar chart shows one bar for each category on the horizontal axis but bar charts can have two or more bars for each category to allow the audience to compare two or more sets of data. When more than one bar is drawn you need to include a key to show which bar is which.

The bar chart below is an example of a dual bar chart showing the number of visitors at a particular destination. A third bar could be added below to show senior citizen visiting figures, if available.

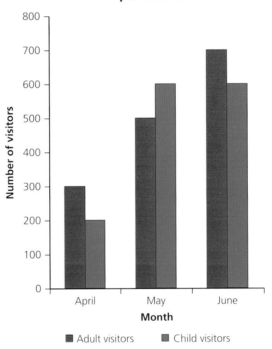

▲ Dual bar chart showing visitor figures to an attraction

Example

You have been asked to produce a visual representation of your organisation's current workforce by age range and gender. Your manager suggests you draw a bar chart.

Solution

Step 1: collate the information and select appropriate age ranges.

Gender and age of employee		
Age	Female	Male
16–24	18	10
25–39	24	30
40–59	14	27
60+	6	11

Step 2: plan or draft your chart and select an appropriate scale for the ages.

Step 3: construct your bar chart, remembering to include a title, a key and labels.

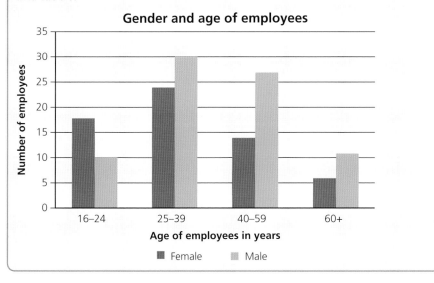

At Stage 2, you need to be able to read and extract information from charts too; you might be given the bar chart from the example above and asked to make observations about the content or identify trends. There will usually be lots of possible observations from any chart. Here are some examples for this chart:

- There are more male employees than female employees.
- There are more employees in the 25–39 age bracket than any other age group.
- There are fewer employees in the 60+ age group than any other age group.

Pictograms

At Stage 2 you need to be able to read and draw pictograms, so let us look at these in more detail.

The most important thing to look at is the **key** because this tells you how many each picture, or image, in the pictogram is worth. At Stage 2 it is unlikely that there will be one image for each item.

An examination question may ask you to total sales for two or more of the rows or it may ask you to compare the totals of different items bought.

Let us look at an example.

Example

The pictogram shows the number of drinks bought from a vending machine on a Monday evening.

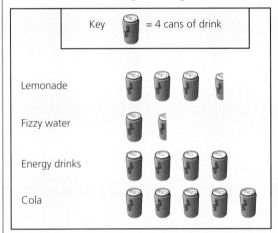

1 How many cans of lemonade and cola were bought from the vending machine?

2 How many more cans of cola than lemonade were bought?

Solution

1 Step 1: look at the key to find out the number of cans each picture represents (4).

Step 2: Find the total number of each type of cans bought

($3\frac{1}{2} \times 4 = 14$ cans of lemonade and $5 \times 4 = 20$ cans of cola)

Step 3: add the numbers for each together ($14 + 20 = 34$ cans)

There were 34 cans of lemonade and cola bought in total.

2 This can be calculated in two ways:

a Work out the number of cans of cola and subtract the number of cans of lemonade.

$20 - 14 = 6$

b Look at the difference.

There are $1\frac{1}{2}$ more cans of cola.

$1\frac{1}{2} \times 4 = 6$

There were 6 more cans of cola bought than cans of lemonade.

Line graphs

Line graphs are usually used to display continuous data and are frequently used by businesses and other organisations to show data about production, sales turnover and profit/loss. A line graph is similar to a bar chart in terms of construction, with vertical and horizontal axes.

The graph shows two line graphs drawn on the same axes showing the average temperature at midnight (blue line) and the average temperature at midday (red line).

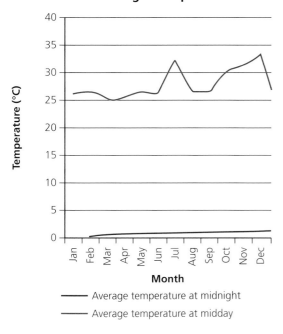

Comparison of average midday and midnight temperatures

—— Average temperature at midnight
—— Average temperature at midday

When you draw a line graph, make sure you label each axis clearly and include a meaningful title. In this example, the axes are labelled *Temperature* (vertical) and *Month* (horizontal) and a suitable title is given at the top.

Multiple line graphs are drawn on the same axes to enable the comparison of two sets of data, for example to compare:

- actual sales or production figures against target sales or production figures
- sales figures for the current year against the previous year
- sales figures for one region against another
- production at one site compared with another.

Tip for assessment

Check the scale carefully when you read numbers from a line graph. For example, the vertical axis may be labelled 'number of people in thousands', meaning when you read 5 it is actually 5 000.

Learner tip

When you draw a graph, make sure you use a scale where you will be able to plot all your data points easily. The scale does not have to start at 0.

Example

You are working for the local tourism board and have been asked to draw a graph to show the average temperature in the UK last year.

Solution

Step 1: collect and organise the required data (calculate average if not already available).

	Jan	Feb	Mar	Apr	May	Jun	Jul	Aug	Sep	Oct	Nov	Dec
UK	3°C	9°C	11°C	15°C	19°C	20°C	24°C	23°C	18°C	14°C	10°C	6°C

Step 2: select an appropriate chart to display findings – a line graph.

Step 3: select an appropriate scale for the temperature. Identify the maximum and minimum data values that you need to plot to help you decide on the scale to use on the temperature axis.

Step 4: draw the axes and include the scale.

Step 5: plot the average temperature for the UK each month and join the plots.

Step 6: label the horizontal axis (Month) and vertical axis (Temperature in degrees Celsius).

Step 7: add an appropriate title explaining the purpose of the graph (Average temperatures in the UK for (year).

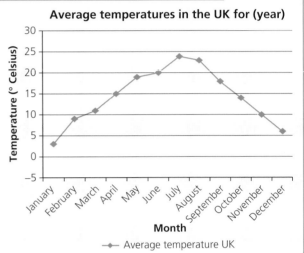

Average temperatures in the UK for (year)

Pie charts

A pie chart displays data as a circle divided into sectors representing the proportions of the total. A pie chart is a great visual tool which can show at a quick glance, for example, that one-third of patients were over 60 or one-quarter of diners ordered dessert C.

Each segment of a pie chart must be labelled clearly, either in the chart or by using a key. Each sector can be labelled as a percentage, as in this example.

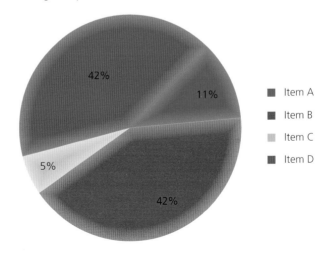

You can use Microsoft Excel to generate pie charts, but it is also useful to be able to draw a pie chart using the traditional tools (a compass, a protractor, a ruler and colouring pencils). You know that there are 360° in a circle and 100% in a whole pie chart.

Therefore, each 1% = 360 ÷ 100 = 3.6°, or 10% = 36°.

Example 1

30% of people at a conference vote to have a shorter lunchtime.

50% vote to keep lunchtime the length it is.

20% do not vote.

How would you represent this on a pie chart?

Solution

Step 1: 50% is straightforward as it is half of the whole or 360 ÷ 2 = 180°.

Step 2: find one per cent by dividing the number of degrees in a circle by the total number of percent: 360 ÷ 100 = 3.6. So 1% = 3.6° and 10% = 36°.

Step 3: find each share.

20% is 2 × 36 or 20 × 3.6 = 72°

30% is 3 × 36 or 30 × 3.6 = 108°

Check the three sectors add up to 360°.

180 + 72 + 108 = 360

Step 4: use a protractor to measure each sector.

Step 5: label and colour each sector.

Step 6: add a title, e.g. Vote on length of lunchtime.

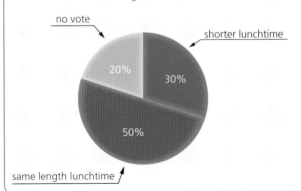

Vote on length of lunchtime

no vote

shorter lunchtime

20%

30%

50%

same length lunchtime

Example 2

Within a group of game console owners, 25 own a PlayStation 3, 40 own an XBox 360 and 35 own a Wii. Draw a pie chart to represent this data.

Solution

Step 1: add the total number of consoles up: 25 + 40 + 35 = 100.

Step 2: find one share by dividing the number of degrees in a circle by the number of consoles: 360 ÷ 100 = 3.6

Step 3: multiply one share by each total to calculate the number of degrees representing each segment:

PlayStation 3.6 × 25 = 90°

XBox 3.6 × 40 = 144°

Wii 3.6 × 35 = 126°

Check that your categories add up to 360°

90° + 144° + 126° = 360°

Step 4: use a protractor to measure each sector.

Step 5: label and colour each sector.

Step 6: add a title, e.g. Breakdown of group console owners

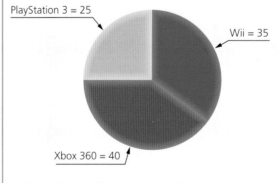

▶ Breakdown of group console owners

A pie chart does not need to include data values, but they have been included here to help you understand the instructions.

The pie chart can then be used to make observations about the data, for example:

- one-quarter of console owners have a PlayStation 3
- three-quarters of console owners have either a Wii or an Xbox 360.

Maps

At Stage 2 you need to be able to read and extract information from a simple map.

Example

A person wishes to travel from A to D. Which is the shortest route?

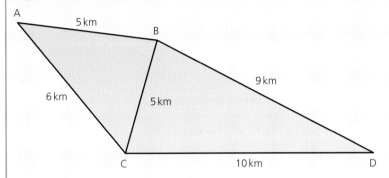

Solution

There are four possible routes:

AB and BD = 5 + 9 = 14 km

AC and CD = 6 + 10 = 16 km

AB, BC and CD = 5 + 5 + 10 = 20 km

AC, CB and BD = 6 + 5 + 9 = 20 km

The shortest route is AB and BD which is 14 km.

Tasks

1 Here are the ages of a group of delegates at a conference.
  Draw a frequency table to show this information.

21	33	27	51	23	26
30	39	23	34	44	49
45	50	38	33	47	34
33	28	20	58	30	30
25	42	29	40	24	35

2 The diagram shows the favourite type of bread in a sample of 200 people.

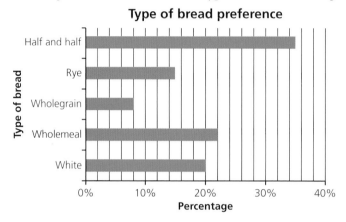

Type of bread preference

  a What percentage of people preferred wholemeal bread?

  b What was the actual number of people preferring white bread?

  c What was the percentage of people whose preference was not half and half?

3 Look at this pictogram.

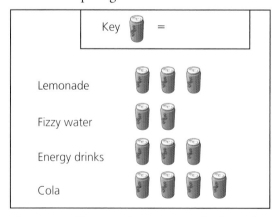

If 24 cans of lemonade were bought from the machine, what number of cans should each picture represent in the key?

4 Look at the pictogram in question 3.
  If 20 cans of cola were bought from the machine, what number of cans should each picture represent in the key?

5 The line graph shows the conversion rate between US Dollar and British Pound (£).

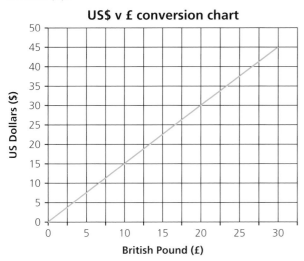

US$ v £ conversion chart

a How many US Dollars would you get if you exchanged £20?

b An item costs 45 US Dollars (US$45). What is the equivalent value in British Pounds (£)?

6 An internet shopping site asks its customers to rate their experience of its service. The results are shown in the table.

Scores on a survey	Number of scores
Excellent	48
Good	105
Average	84
Poor	39
Very poor	24

a Present the results in a pie chart.

b What observations can you make from the pie chart?

1 The graph shows average sales of beachwear in a department store throughout the year.

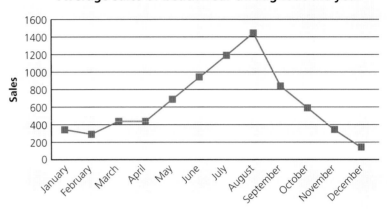

Average sales of beachwear throughout the year

Which of the following conclusions can you draw from the chart?

a Sales of beachwear are highest in January.

b There is a steady increase in sales as summer approaches followed by a decline towards Christmas.

c There is no real trend in sales.

d Most people buy beachwear in Spring.

2 The pie chart shows the television viewers in the UK by channel.

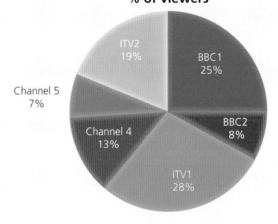

% of viewers

Which channel gets a quarter of all viewers?

a BBC1

b ITV1

c ITV2

d Channel 4

3 You want to book a week's holiday, half board, starting on 3rd May for a family of three.

The table shows the prices for your chosen resort.

Rhodes Village ****	1st April–30th May	1st June–20 July	21st July–7 September	8th September–31 October
	Per Person Per Night ($)	Per Person Per Night ($)	Per Person Per Night (£$)	Per Person Per Night ($)
Room only	30	45	70	55
Bed + breakfast	50	60	85	70
Half board	80	90	100	85
All inclusive	95	110	135	120

How much will the holiday cost?

a $240

b $270

c $560

d $1680

4 The bar chart shows the results of a survey investigating people's favourite takeaway food.

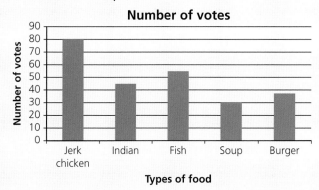

How many more people voted for Jerk chicken than Soup?

a 25

b 35

c 50

d 80

5 The line graph shows a new supermarket's sales during its first year of trading.

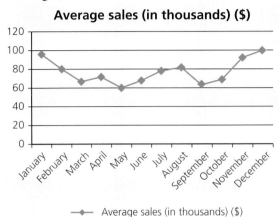

What was the difference between the highest and lowest monthly sales figures?

a $4 000

b $40 000

c $6 000

d $60 000

6 The bar chart shows an organisation's employees by age and gender.

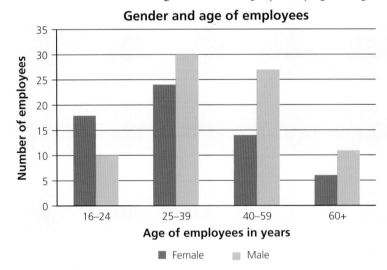

Which age group has the least difference, in number, between male and female employees?

a 16–25

b 26–40

c 41–60

d over 60

7 The manager at a swimming pool wants to display the number of visitors to the pool over one week. The assistant manager has produced the following charts and diagrams to show the data. Which chart displays the data fully?

a Number of swimmers

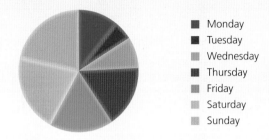
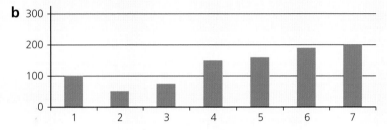

b

c Number of swimmers

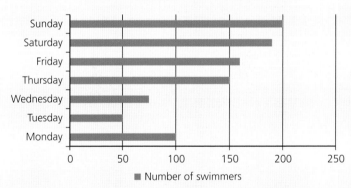

d Number of swimmers

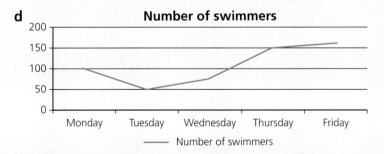

8 The table shows the number of customers at each attraction at a fairground over a weekend.

Attraction	Number of customers
Rollercoaster	1068
Waltzers	470
Pirate Ship	672
Ghost Train	268
Fun House	209

They decide to present this information using two visual displays, one showing the number of visitors per attraction and one showing the proportion per attraction.

Which combination of diagrams would best suit their needs?

a Tally chart and bar chart

b Bar chart and pie chart

c Bar chart and line graph

d Pictogram and line graph

9 The pictogram shows the number of cars sold, by colour, in January.

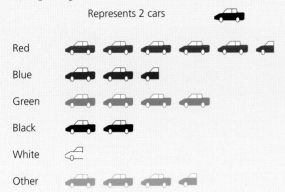

Represents 2 cars

Red

Blue

Green

Black

White

Other

How many red, white and blue cars were sold in January?

a 10

b 17

c 20

d 23

10 25% of passengers on a cruise ship were American.
How many degrees would represent this on a pie chart?

a 25

b 45

c 90

d 180

STAGE 3

Unit 301
Operations on integers

Introduction

At this stage you will be confident in adding, subtracting, multiplying and dividing whole numbers of any size. You also know that division is the inverse of multiplication. Even though you may never be without the calculator on your mobile phone, it is still very important to be able to add and subtract quickly in your head and on paper. You need to know if your calculator gives you the wrong answer, for example if you have made a mistake in typing in the calculation and, with practice, it is often quicker to add or subtract in your head. Also, you will not be able to use a calculator in the City & Guilds' examination.

Learning objectives

This unit will introduce the mathematical terms:

- prime numbers
- multiples
- factors
- natural numbers
- integers and positive integers.

The unit will also cover:

- squares and cubes of positive integers
- square roots
- standard form
- directed numbers
- other number systems such as binary.

These are all topics that you may not come across in everyday life but will help you to solve problems and to progress further with your studies in mathematics.

What is an integer?

Integers are whole numbers.

Positive integers are 1, 2, 3, 4 and so on.

Operations on whole numbers: addition, subtraction, multiplication and division

In Unit 201, we looked at **place value** and included millions. When working with big numbers, it is helpful to use a formal written method with columns. The most important thing to remember is to make sure the digits line up: units with the units, tens with the tens and hundreds with the hundreds. If they are not lined up you may confuse a hundred for ten and get the wrong answer. You can use any method that works for you. You may find it useful to review Unit 203 before you try these.

Addition

1 A property costs four hundred and fifty thousand dollars.
   The alterations are estimated to cost $73 450. What is the total cost?

Subtraction

2 Another property costs one and a half million dollars. A purchaser has $350 000.
   How much does he need to borrow?

Multiplication

3 A customer pays $125 per month on credit payments for a car.
   What is the total amount he pays over 3 years?

Division

4 A grandmother shares $45 000 between her 12 grandchildren.
   How much does each grandchild receive?

Answers

1 $523 450

2 $1 150 000

3 $4 500

4 $3 750

Did you get all four answers correct? Well done!

If not, review Unit 201 on Place value and Unit 203, Operations on whole numbers, before you continue.

Multiples

For multiples, you need to remember your times tables.

A **multiple** is a number that may be divided by another number a certain number of times without a remainder.

The multiples of 2 are all the numbers in the 2 times table, such as 2, 4, 6, 8, 10 and so on, but they include really big numbers as well: 200 000 is a multiple of 2.

Multiples of 2 always end with a 2, 4, 6, 8 or 0.

The multiples of 5 are all the numbers in the 5 times table, such as 5, 10, 15, 20, 25 and so on but they include really big numbers as well: 200 000 is a multiple of 5.

Multiples of 5 always end with a 5 or a 0.

> **Example 1**
>
> Is 18 a multiple of 3?
>
> Solution
>
> Yes. The multiples of 3 are as in the three times table: 3, 6, 9, 12, 15, 18 and so on.

Example 2

List the first three multiples of 7.

Solution

The multiples of 7 are as in the seven times table: 7, 14, 21, 28 and so on.

The first three multiples are 7, 14 and 21.

Example 3

Do 5 and 3 have any multiples the same?

Solution

Yes. The multiples of 5 are as in the five times table: 5, 10, 15, 20, 25 and so on. The multiples of 3 are as in the three times table: 3, 6, 9, 12, 15, 18 and so on.

We can see that the number 15 appears in both lists. Therefore, 15 is a multiple of 3 and a multiple of 5.

This is obvious as $5 \times 3 = 15$. We can double 15 and know that 30 is also a multiple of 5 and 3.

Factors

Real world maths

We use factors all time. For example, if a classmate asks you to exchange $500 for smaller notes, you will have to think about the number of $50 and/or $100 notes that will be equivalent to $500.

Can you think of other times that you use factoring? What about sharing things equally, calculating the time it takes to travel from one place to the next, or comparing the prices of similar items?

Factors are numbers that divide exactly into another number.

For example, the factors of 6 are:

1, 2, 3, 6

Factors can be shown in pairs. Each pair multiplies to make 6.

The factor pairs of 6 are

$1 \times 6 = 6$

$2 \times 3 = 6$

The factors of 8 are:

1, 2, 4, 8

The factor pairs of 8 are

$1 \times 8 = 8$

$2 \times 4 = 8$

Example 1

What are the factors of 10?

Solution

The numbers that divide exactly into 10 are the factors. These are 1, 2, 5, 10

The factor pairs are:

1×10

2×5

Example 2

What are the factors of 4?

Solution

The numbers that divide exactly into 4 are the factors. These are 1, 2, 4

The factor pairs are:

1×4

2×2

Example 3

What are the factors of 7?

Solution

The numbers that divide exactly into 7 are the factors. These are 1, 7

The factor pairs are:

1×7

No other numbers divide exactly into 7, so 7 is a prime number – see below.

Prime numbers

Prime numbers are special numbers that can only be divided by themselves and 1. The number 1 is not counted as a prime number.

7 is a prime number. It can only be divided by 1 and 7. We can also say the factors of 7 are 1 and 7.

9 is not a prime number. The factors of 9 are 1, 3 and 9. Therefore, 9 is not a prime number as it can be divided by 3 as well as by 1 and 9.

Example 1

Is 10 a prime number?

Solution

The factors of 10 are:

1, 2, 5, 10

So 10 is not a prime number.

Example 2

Is 4 a prime number?

Solution

The factors of 4 are:

1, 2, 4

So 4 is not a prime number.

Activity

Find all the prime numbers below 20.

Example 3

Is 17 a prime number?

Solution

The factors of 17 are:

1, 17

So 17 is a prime number.

Natural numbers

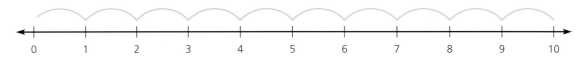

Natural numbers are those used for counting.

They are positive numbers: 1, 2, 3, 4 and so on.

Square numbers

We know that a square has four equal sides.

When we found the area of a square $2\,\text{m} \times 2\,\text{m}$, it was $4\,\text{m}^2$. We say this as four metres squared or four square metres.

A **square number** is a number multiplied by itself. This can also be called 'a number squared'. The symbol for squared is ^{2}.

$2^2 = 2 \times 2 = 4$

$5^2 = 5 \times 5 = 25$

Example 1

What is the value of 3^2?

Solution

$3^2 = 3 \times 3 = 9$

Example 2

What is the value of 10^2?

Solution

$10^2 = 10 \times 10 = 100$

Example 3

What is the value of 7^2?

Solution

$7^2 = 7 \times 7 = 49$

Activity

Find all the square numbers with a value of up to 100.

Square roots

When we found the area of a square 2 m × 2 m, it was 4 m². If you know that the area of a square is 4 m², you can work out the sides of the square by finding the **square root**. We use the symbol $\sqrt{\ }$ to signify square root.

$\sqrt{4} = 2$

> ### Example 1
> What is the value of $\sqrt{100}$?
>
> ### Solution
> From your tables, you know 10 × 10 = 100. Or you can use the $\sqrt{\ }$ button on your calculator. So $\sqrt{100} = 10$

> ### Example 2
> What is the value of $\sqrt{36}$?
>
> ### Solution
> From your tables, you know 6 × 6 = 36. So $\sqrt{36} = 6$

> ### Tip for assessment
> For the City & Guilds exam, you will not be able to use a calculator but the questions will only include whole numbers up to 12 (square roots up to 144).

Cube numbers

We know that a cube is a three-dimensional shape with equal sides. When we found the volume of a cube 2 m × 2 m × 2 m, it was 8 m³. We say this as eight metres cubed or eight cubic metres.

A **cube number** is a number multiplied by itself 3 times. This can also be called 'a number cubed'. The symbol for cubed is ³.

$2^3 = 2 \times 2 \times 2 = 8$

$4^3 = 4 \times 4 \times 4 = 64$

> ### Example 1
> What is the value of 3^3?
>
> ### Solution
> $3^3 = 3 \times 3 \times 3 = 27$

> ### Example 2
> What is the value of 10^3?
>
> ### Solution
> $10^3 = 10 \times 10 \times 10 = 1000$

> ### Example 3
> What is the value of 5^3?
>
> ### Solution
> $5^3 = 5 \times 5 \times 5 = 125$

> ### Activity
> Find all the cube numbers with a value of up to 100.

Standard form

Standard form is a way of writing down very large or very small numbers easily.

Standard form is also sometimes referred to as scientific notation because scientists use standard form when working with large numbers such as the speed of light or small numbers such as the size of atoms.

The system is based on using powers of 10 to express how big or small a number is.

We know that $10^2 = 10 \times 10 = 100$ and $10^3 = 10 \times 10 \times 10 = 1000$. These are known as 10 to the power of 2 and 10 to the power of 3, respectively.

The values of different powers of 10 are shown here.

$10^0 = 1$

$10^1 = 10$

$10^2 = 100$

$10^3 = 1\,000$

$10^4 = 10\,000$

$10^5 = 100\,000$

$10^6 = 1\,000\,000$

So $20\,000 = 2 \times 10\,000$

$10\,000 = 10 \times 10 \times 10 \times 10 = 10^4$

Therefore $20\,000 = 2 \times 10^4$ in standard form.

Standard form is always written as a number (from 1 to 9.9) $\times$ 10 to the relevant power.

Example 1

Write $50\,000$ in standard form.

Solution

$50\,000$ can be written as $5 \times 10\,000$

$10\,000 = 10 \times 10 \times 10 \times 10 = 10^4$

So $50\,000 = 5 \times 10^4$ in standard form.

Example 2

Write $6\,129$ in standard form.

Solution

$6\,129$ can be written as $6.129 \times 1\,000$

$1\,000 = 10 \times 10 \times 10 = 10^3$

So $6\,129 = 6.129 \times 10^3$ in standard form.

Example 3

Write 7 050 000 in standard form.

Solution

7 050 000 can be written as $7.05 \times 1\,000\,000$

$1\,000\,000 = 10 \times 10 \times 10 \times 10 \times 10 \times 10 = 10^6$

So $7\,050\,000 = 7.05 \times 10^6$ in standard form.

Example 4

Write 4.9×10^3 as an integer.

Solution

$10^3 = 10 \times 10 \times 10 = 1\,000$

So this is $4.9 \times 1\,000 = 4\,900$

Directed numbers

Directed numbers can be positive or negative.

Ten is a **positive number**. If you count down from ten to zero, what happens if you want to carry on counting down?

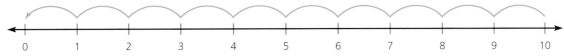

We can, but we use a minus symbol in front of the number to show it's below zero, and say negative 1 or minus 1, negative 2, negative 3, and so on.

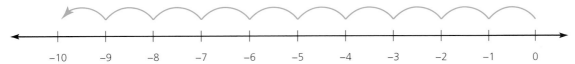

Remember, the negative sign is part of the number and does not mean you have to take away. When looking at a number, look first to see if there is a negative (minus) sign.

Think what this means: −6 is lower down the number line than −2. If it was temperature, −6 °C would be colder than −2 °C.

Negative numbers have many uses in everyday life, for example to show money withdrawal from a bank account, to indicate temperatures below zero and to show height below sea level.

Example 1

The temperature of a refrigerator is 4 °C and the temperature of a freezer is −18 °C.

a Which is the coldest temperature?

b What is the difference in temperatures?

Solution

a Think of the number line. Negative 18 is below zero, so −18 °C is colder than 4 °C.

b Think of the number line. Negative 18 is 18 below zero and 4 °C is 4 above zero, so the difference is 18 and 4 which is 22 °C.

Example 2

A customer has $100 in his account. He makes a payment for $110.

What is the balance on his account?

Solution

Think of the number line. Take away 100 and you still need to take away another $10, so this is negative 10. The balance on the account is −$10 or $10 overdrawn.

Time differences are sometimes shown as negative when one time zone is behind another time zone. You may have watched the news coverage for New Year. Sydney, Australia celebrates New Year 16 hours ahead of Kingston, Jamaica.

Example 3

What time is it in Jamaica when Sydney, Australia celebrates New Year?

Solution

The difference is −16 so count back 16 from 24. This is the same as
24 − 16 = 8

The time is 08:00 or 8 a.m.

▲ Sydney

Number systems

The standard number system used around the world is base 10. This is also known as a **decimal number system** or denary system. It uses ten digits (0, 1, 2, 3, 4, 5, 6, 7, 8, and 9) to represent all numbers. We count to 9 in the units column and then when we add 1 we have one ten and no units. Similarly, when we have 9 tens and 9 units (99) we add 1 and get one hundred, no tens and no units (100). See Unit 201 on Place value.

What is this number?

256.9

Million	Hundred thousand	Ten thousand	Thousand	Hundred	Ten	Unit	.	1/10	1/100	1/1000
				2	5	6		9		

This number reads, two hundred and fifty-six point nine.

It means two hundreds, five tens, six units and nine tenths.

Binary

The standard number system used by computers is the **binary system** or base 2. For binary numbers there are only two possible digits available: 0 or 1. These are called **bits**. You may have heard of the term bytes used in computing. A byte is a group of 8 bits but you will not be asked to covert with bytes at Stage 2 (a kilobyte (KB) is 1 000 bytes).

The binary system has place value columns. The columns are arranged in multiples of 2:

Sixteens $2 \times 2 \times 2 \times 2$	Eights $2 \times 2 \times 2$	Fours 2×2	Twos 2	Units Ones
1	0	0		1

This binary number reads 'one zero zero one' or one eights, zero fours, zero twos and one unit. It does not read 'one thousand and one' because it is not in the denary system.

By looking at the place values we can calculate the equivalent denary number:

one × eight, zero fours, zero twos and one unit

$$8 \quad + \quad 0 \quad + \quad 0 \quad + \quad 1$$

This equals 9. So 1001 in base 2 = 9 in base 10

We can also convert to binary from decimal.

What is 7 in base 10 written as a binary number?

Sixteens $2 \times 2 \times 2 \times 2$	Eights $2 \times 2 \times 2$	Fours 2×2	Twos 2	Units Ones
0	0	1	1	1

There are no eights, but one 4 so we put 1 in the fours column. We take 4 from 7 and are left with 3. There is one 2 in 3, so we put 1 in the twos column. There is 1 left, so we put 1 in the ones column.

7 in base 10 is 111 written as a binary number.

Example 1

What is the binary number 101 in the decimal (denary) system?

Solution

Sixteens $2 \times 2 \times 2 \times 2$	Eights $2 \times 2 \times 2$	Fours 2×2	Twos 2	Units Ones
0	0	1	0	1

The number is one × four, zero twos and one unit.

$4 + 0 + 1 = 5$

The binary number 101 is 5 in the decimal system.

Example 2

What is the decimal number 18 in the binary system?

Solution

Sixteens $2 \times 2 \times 2 \times 2$	Eights $2 \times 2 \times 2$	Fours 2×2	Twos 2	Units Ones
1	0	0	1	0

There is one × sixteen and two left. So zero × eights and zero × fours but one × two. No units.

The decimal number 18 is 10010 in the binary system.

Example 3

What is the binary number 1010 in the decimal (denary) system?

Solution

Sixteens $2 \times 2 \times 2 \times 2$	Eights $2 \times 2 \times 2$	Fours 2×2	Twos 2	Units Ones
0	1	0	1	0

The number is one × eight, zero fours, one × two and no units.

$8 + 0 + 2 + 0 = 10$

The binary number 1010 is 10 in the decimal system.

Tasks

1 The table shows heights above sea level of some features in different African countries.

Chott Melrhir, Algeria	Limpopo & Shashe Rivers, Botswana	Danakil, Ethiopia	Sabkhat Ghuzayyil, Libya	Niger River, Niger
−40 m	513 m	−125 m	−47 m	200 m

    a Put these heights in order from lowest to highest.

    b What is the difference between the lowest and the highest heights?

2 Order these temperatures from coldest to hottest.

    267 °C   −78 °C   19 °C   38 °C   −16 °C

3 Which of these numbers are prime numbers?

    6   9   11   13   15

4 **a** What is 4^2?

    b What is 2^3?

5 List the multiples of 3 up to 20.

6 What are the factors of 12?

7 Write 16 000 in standard form.

8 What is the binary number 111 as a denary number?

9 A man saves $256 a month for six years. How much does he save?

10 What is an integer?

▶ The Limpopo River

Test your knowledge

In your exam, the questions are multiple choice. This means that you need to choose the correct answer from the four options given: a, b, c and d. The correct answer will always be shown, but the other answers often look reasonable and may be the result of a simple mistake or an error in the method.

For example, which of these numbers has 6 hundreds?

    a 286

    b 682

    c 268

    d 862

All the options have a 6 in them, but only one has 6 in the hundreds column so (b) is the correct answer.

Try these

1 Which of the following sets of numbers contains all of the factors of 20?

    a 1, 2, 10, 20

    b 1, 4, 8, 12, 16, 20

    c 1, 2, 4, 5, 10, 20

    d 2, 4, 6, 8, 10, 12, 14, 16, 18, 20

2 $\sqrt{64} =$

    a 8

    b 16

    c 32

    d 128

3 What is 17 000 written in standard form?

    a 0.17×10^3

    b 1.7×10^3

    c 1.7×10^4

    d 17×10^5

4 Order these temperatures from coldest to hottest.

    7°C   −12°C   19°C   32°C   −1°C

a −1°C	7°C	−12°C	19°C	32°C
b −1°C	−12°C	7°C	19°C	32°C
c −12°C	−1°C	7°C	19°C	32°C
d 32°C	19°C	7°C	−1°C	−12°C

5 What is 2.6×10^2 written as an integer?

   a 0.26

   b 26

   c 260

   d 2600

6 Which of these numbers is a prime number?

   a 4

   b 9

   c 15

   d 23

7 What is 7^3?

   a 14

   b 21

   c 49

   d 343

8 Which of these numbers is a multiple of 4?

   a 0

   b 1

   c 2

   d 8

9 What is $44\,400 + 3\,075 - 12\,678$?

   a 32 347

   b 34 797

   c 34 803

   d 35 203

10 What is $3^2 + \sqrt{25}$?

   a 8

   b 14

   c 28

   d 34

Unit 302
Operations on decimal fractions

Introduction

Mathematics has three ways to describe parts of numbers: using a **fraction**, a **decimal** or a **percentage**. This unit will cover addition, subtraction, multiplication and division of decimal fractions.

This unit builds on the Stage 2 topic on decimal fractions (Unit 204), where you multiplied and divided decimal fractions by whole numbers. You may like to remind yourself of the unit.

Learning objectives

In this unit you will find information on:

- adding and subtracting two or more numbers with no more than three decimal places
- multiplying a number with up to four digits and no more than two decimal places by a decimal number
- dividing a number with not more than four digits and two decimal places by a decimal number
- solving problems using a combination of operations.

You will find decimal fractions useful when working with measures in the decimal system: kilograms (for weight), metres (for distance and length) and litres (for capacity) are all based on tens.

Addition and subtraction of decimal fractions

You can add and subtract decimals in the same way as whole numbers. There are various different methods you can use. Choose the method, or methods, you find easiest to understand.

Addition and subtraction of decimal numbers

Make sure you line up the decimal points and the tenths and hundredths and thousandths after it as well as the hundreds, tens and units before it.

Example 1

Work out $10 + 0.35 + 0.2$

Solution

```
   10.00
 +  0.35
 +  0.2
 ─────────
   10.55
```

First, set the addition in columns. Remember 10 can be written as 10.00. 0.2 can be written as 0.20, if you wish.

Add the digits in the hundredths column ($0 + 5 + 0 = 5$, write 5 in the hundredths column).

Add the digits in the tenths column ($0 + 3 + 2 = 5$, write the 5 in the tenths column).

Add the digits in the units column ($0 + 0 + 0 = 0$, write the 0 in the units column).

There is only 1 digit in the tens column, so write this (1).

$10 + 0.35 + 0.2 = 10.55$

Example 2

Use column subtraction to work out $10 - 0.24$.

Solution

```
   ⁰9̶1̶0̶ . ⁹1̶0̶ ¹0
 −   0 .  2  4
 ─────────────────
     9 .  7  6
```

First set the subtraction in columns.

Subtract the digits in the hundredths column. ($0 - 4$ can't be done, so borrow a ten. $10 - 4 = 6$, write 6 in the hundredths column.)

Subtract the digits in the tenths column. You have borrowed 10 so you only have 9 in this column. ($9 - 2 = 7$, write the 7 in the tenths column.)

Subtract the digits in the units column. You have borrowed 10 so you only have 9 in this column. ($9 - 0 = 9$, write the 9 in the tenths column.)

In the tens column, you have borrowed 10 so you have 0 in this column.

$10 - 0.24 = 9.76$

Multiplying a decimal by a decimal

$$
\begin{array}{r}
4.90 \\
\times\ 0.3 \\
\hline
1.470
\end{array}
$$

To multiply a decimal by a decimal, you do not need to line up your work as carefully as you will count the decimal points in the question and the answer.

Put the first digit you work out under the column on the right. There are 3 digits after the decimal point in the question so count three digits from the right-hand side in the answer.

$4.90 \times 0.3 = 1.470$ or 1.47

Compare with this

$$
\begin{array}{r}
4.9 \\
\times\ 0.3 \\
\hline
1.47
\end{array}
$$

Put the first digit you work out under the column on the right. There are 2 digits after the decimal point in the question so count two digits from the right-hand side in the answer.

Both calculations are working with the same amounts. 4.90 is the same as 4.9 and 1.470 is the same as 1.47.

You can do a rough estimate to check your answer: $5 \times 0.3 = 1.5$

0.3 is less than 1 so the answer will be less than 4.9.

Example 1

41.25×1.3

Solution

Remember to set out the sum.

$$
\begin{array}{r}
41.25 \\
\times\ 1.3 \\
\hline
12375 \\
41250 \\
\hline
53625
\end{array}
$$

Put the first digit you work out under the column on the right. There are 3 digits after the decimal point in the question so count three digits from the right-hand side in the answer. This gives 53.625

You can do a rough estimate to check your answer, $40 \times 1 = 40$

Example 2

0.382×0.72

Solution

```
  0.382
× 0.72
─────
  0764
 26740
─────
 27504
```

There are 5 digits after the decimal point in the question so count five digits from the right-hand side in the answer.

This gives 0.27504

> You can do a rough estimate to check your answer, $0.3 \times 0.7 = 0.21$
>
> Both numbers are less than 1 so there will be no units.

> **Learner tip**
>
> Make a rough estimate of the answer.
>
> $4.1 \times 0.6 = 2.46$. This could be $4 \times 1 = 4$ or $4 \times 0.5 = 2$. So the answer will not be 246 or 0.246

You can also count the number of digits after the decimal point in the question and check there is the same number of digits after the decimal point in the answer.

Dividing a decimal by a decimal

Set out the division sum as normal.

If you are dividing by a decimal number rather than a whole number, move the decimal point to the right to make it a whole number and move the decimal point in the number you are dividing (the dividend) the same number of places.

Put the decimal point on the answer line directly above the decimal point in the dividend.

Then divide as usual.

If the calculation is divide 320.84 by 0.4, the calculation would look like this:

$0.4\overline{)320.84}$

To make 0.4 a whole number, the decimal point needs to move 1 digit to the right and you need to move the decimal point in 320.84 by 1 digit.

This gives

$4\overline{)3208.4}$

Now divide as usual

$$\begin{array}{r} 802.1 \\ 4\overline{)3208.4} \end{array}$$

> Remember, line up the decimal point on the answer line above the decimal point in the number before you start to divide.

Example 1

$1.95 \div 0.13$

Solution

This becomes $195 \div 13$ when you move the decimal place two digits to the right to make the divisor a whole number.

13 does not go into 1, try 19 – goes once with 6 left over. Bring down the 5 so it is 13 into 65.

$5 \times 13 = 65$, so $1.95 \div 0.13 = 15$

The answer is 15

Example 2

7202.76 ÷ 0.3

Solution

This becomes 72 027.6 ÷ 3 when you move the decimal place one digit to the right to make the divisor a whole number.

3 goes into 7 twice (so put 2 on the answer line) with 1 left over. Bring down the 2 so 3 into 12 goes 4 (put 4 on the answer line). Bring down the 0, 3 does not go into 0 (so put 0 on the answer line). 3 into 2 does not go (so put 0 on the answer line). Bring down 7. 3 into 27 goes 9 times (put 9 on the answer line). 3 into 6 goes twice so put 2 on the answer line after the decimal point.

7202.76 ÷ 0.3 = 24 009.2

The answer is 24 009.2

Tasks

1 Add 1.6, 3.05 and 24.029

2 2 m + 13.05 m + 2.345 m

3 12 − 0.055

4 706.5 − 2.099

5 0.25 × 2.9

6 107.8 × 0.552

7 247.8 ÷ 0.25

8 0.249 ÷ 0.3

9 How many pieces of wood 0.7 m long can be cut from a 3.75 m length of wood?

10 Find the total cost of 1.5 kg cheese at $6 per kg and 0.25 kg of butter at $8 per kg.

1 What is 0.3×0.7?

   a 0.201   **c** 2.1

   b 0.21   **d** 21

2 What is $7.34 + 10.7 + 0.6$?

   a 8.47   **c** 17.64

   b 17.47   **d** 18.64

3 What is $2.1 - 0.735$?

   a 1.365   **c** 2.365

   b 1.435   **d** 2.635

4 What is 14.5×0.12?

   a 1.63   **c** 16.3

   b 1.74   **d** 174

5 What is 0.28×0.2?

   a 0.056   **c** 5.6

   b 0.56   **d** 56

6 What is $30.24 \div 0.8$?

   a 3.03   **c** 3.88

   b 3.08   **d** 37.8

7 What is $10.5 \div 0.21$?

   a 0.5   **c** 5

   b 2   **d** 50

8 Find the total cost of 2.5 kg brown sugar at $1.90 per kg and 0.5 kg of cheese at $18.60 per kg.

   a $14.05   **c** $23.50

   b $20.50   **d** $61.50

9 A piece of ribbon is 2.75 m long. How many pieces 0.25 m long can be cut from this?

   a 11   **c** 9

   b 10   **d** 6

10 What is 0.4×0.04?

   a 0.0016   **c** 0.16

   b 0.016   **d** 1.6

Unit 303
Operations on common fractions

Introduction

Mathematics has three ways to describe parts of numbers: using a **fraction**, a **decimal** or a **percentage**. This unit will cover addition, subtraction, multiplication and division of common fractions.

This unit builds on Unit 205, which introduced addition and subtraction of fractions. You may like to remind yourself of the unit.

In this unit you will find information on:

- adding and subtracting fractions, including borrowing
- multiplying fractions
- dividing fractions
- solving problems using a combination of operations.

Adding fractions

To add fractions, all the fractions must be the same size, such as all halves or all quarters.

If we add two-quarters and one-quarter, we get three-quarters.

$$\frac{2}{4} + \frac{1}{4} = \frac{3}{4}$$

We can add these across the numerators (the top number) because this tells us how many parts we have. But we can only do this if the denominators (the bottom number) are the same. Note we do not add the bottom numbers.

We can also write this as $\frac{2 + 1 = 3}{4}$

What is $1\frac{3}{5} + 2\frac{3}{10}$?

Here the denominator is not the same, so we cannot simply add the top numbers.

Think of the equivalent fractions section in Unit 205. How can we change the denominators so that they are the same?

$\frac{1}{5} = \frac{2}{10}$ ⟵

> Look through the equivalencies section again, if you need to.

If $\frac{1}{5}$ is $\frac{2}{10}$, then $\frac{3}{10}$ must be 3 lots of $\frac{2}{10}$, so $\frac{2}{10} + \frac{2}{10} + \frac{2}{10}$ which is $\frac{6}{10}$

Now we can add the fractions as the denominators are the same:

$\frac{6}{10} + \frac{3}{10} = \frac{9}{10}$

Then add the whole numbers $1 + 2 = 3$

So, the answer is $3\frac{9}{10}$

You could change the whole numbers to tenths but there is no need to do this.

What is $3\frac{2}{3} + \frac{2}{5}$?

We cannot change $\frac{2}{3}$ to fifths so we need to find a number that both 3 and 5 divide into. Think of the multiples section in Unit 301.

Multiples of 3: 3, 6, 9, 12,⑮, 18 and so on.

Multiples of 5: 5, 10,⑮, 20 and so on.

We can see that 15 is a multiple of both 3 and 5 so we can change each fraction to $\frac{?}{15}$

> **Learner tip**
> Remember you can multiply the two numbers together to find a common denominator.

$\frac{2}{3} = \frac{?}{15}$

$\frac{2 \times 5 = 10}{3 \times 5 = 15}$ ⟵

> $3 \times 5 = 15$, so multiply the numerator by 5.
> $2 \times 5 = 10$

$\frac{2 \times 3 = 6}{5 \times 3 = 15}$ ⟵

> $5 \times 3 = 15$, so multiply the numerator by 3.
> $2 \times 3 = 6$

$\frac{10 + 6}{15} = \frac{16}{15} = 1\frac{1}{15}$

$1\frac{1}{15} + 3 = 4\frac{1}{15}$

Example

a $\dfrac{7}{10} + \dfrac{3}{15}$

b $2\dfrac{3}{4} + \dfrac{1}{3}$

c $2\dfrac{1}{2} + 7\dfrac{3}{5}$

Solution

a Find a common denominator.

Both 10 and 15 divide into 30 (you could use 60, 90, 120 or 150 but it is easiest to use the lowest number).

$$\dfrac{7}{10} = \dfrac{7 \times 3}{10 \times 3} = \dfrac{21}{30}$$

$$\dfrac{3}{15} = \dfrac{3 \times 2}{15 \times 2} = \dfrac{6}{30}$$

so $\dfrac{21 + 6}{30} = \dfrac{27}{30}$

b Find a common denominator.

Both 4 and 3 divide into 12.

Keep the whole number separate.

$$\dfrac{3}{4} = \dfrac{3 \times 3}{4 \times 3} = \dfrac{9}{12}$$

$$\dfrac{1}{3} = \dfrac{1 \times 4}{3 \times 4} = \dfrac{4}{12}$$

so $\dfrac{9 + 4}{12} = \dfrac{13}{12} = 1\dfrac{1}{12}$ then add the whole number again $+\, 2 = 3\dfrac{1}{12}$

c Find a common denominator.

Both 2 and 5 divide into 10.

Keep the whole number separate.

$$\dfrac{1}{2} = \dfrac{5}{10}$$

$$\dfrac{3}{5} = \dfrac{6}{10}$$

so $\dfrac{5 + 6}{10} = \dfrac{11}{10} = 1\dfrac{1}{10}$ then add the whole numbers $+\, 2 + 7 = 10\dfrac{1}{10}$

Subtracting fractions

To subtract fractions, all the fractions must be the same size, such as all halves or all quarters.

What is $\dfrac{1}{2} - \dfrac{1}{4}$?

To subtract fractions, all the fractions must be the same size. A quarter is not as big as a half so we need to make the half into quarters.

$$\dfrac{1}{2} = \dfrac{2}{4}$$

Now we can do the subtraction sum $\dfrac{2}{4} - \dfrac{1}{4} = \dfrac{1}{4}$

We can also write this as $\dfrac{2-1}{4} = \dfrac{1}{4}$

What is $3\dfrac{2}{3} - \dfrac{2}{5}$?

We cannot change $\dfrac{2}{3}$ to fifths so we need to find a number that both 3 and

5 divide into. Think of the **multiples** section in Unit 301.

Multiples of 3: 3, 6, 9, 12, ⑮, 18 and so on.

Multiples of 5: 5, 10, ⑮, 20 and so on.

We can see that 15 is a multiple of both 3 and 5 so we can change each

fraction to $\dfrac{?}{15}$

$\dfrac{2}{3} = \dfrac{?}{15}$

$\dfrac{2 \times 5}{3 \times 5} = \dfrac{10}{15}$

> $3 \times 5 = 15$, so multiply the numerator by 5.
>
> $2 \times 5 = 10$

$\dfrac{2 \times 3 = 6}{5 \times 3 = 15}$

> $5 \times 3 = 15$, so multiply the numerator by 3.
>
> $2 \times 3 = 6$

$\dfrac{10 - 6}{15} = \dfrac{4}{15}$

So $\dfrac{4}{15} + 3 = 3\dfrac{4}{15}$

Example

a $\dfrac{3}{4} - \dfrac{3}{5}$

b $2\dfrac{1}{2} - 1\dfrac{5}{8}$

c $3\dfrac{1}{2} - 1\dfrac{2}{3}$

Solution

a A common denominator is 20:

$\dfrac{3}{4} = \dfrac{15}{20}$ and $\dfrac{3}{5} = \dfrac{12}{20}$

$\dfrac{15 - 12}{20} = \dfrac{3}{20}$

b A common denominator is 8:

$\dfrac{1}{2} = \dfrac{4}{8}$

You can't take $\dfrac{5}{8}$ from $\dfrac{4}{8}$ so change 1 whole to eighths.

$1\dfrac{12}{8} - 1\dfrac{5}{8}$

$\dfrac{12 - 5}{8} = \dfrac{7}{8}$ and $1 - 1 = 0$

so the answer is $\dfrac{7}{8}$

c A common denominator is 6:

$3\dfrac{3}{6} - 1\dfrac{4}{6}$

You can't take $\dfrac{4}{6}$ from $\dfrac{3}{6}$ so change 1 whole to sixths:

$2\dfrac{9}{6} - 1\dfrac{4}{6} = 1\dfrac{5}{6}$

Multiplying fractions

To multiply fractions we multiply the numerators of the fractions to get the new numerator. Then multiply the denominators of the fractions to get the new denominator. Simplify the resulting fraction as necessary.

$$\frac{1}{2} \times \frac{3}{4} = \frac{1 \times 3}{2 \times 4} = \frac{3}{8}$$

You can check this by thinking about the pizza cut into 8 slices. You have $\frac{3}{4}$ left which is $\frac{6}{8}$ or six slices. Half of 6 is 3 so you have $\frac{3}{8}$ left.

$$2\frac{1}{2} \times \frac{3}{4}$$

If you have a whole number, you need to make this into a fraction.

$$2\frac{1}{2} = \frac{5}{2}$$

$$\frac{5 \times 3}{2 \times 4} = \frac{15}{8} = 1\frac{7}{8}$$

Example

a $\frac{3}{4} \times \frac{3}{5}$

b $1\frac{1}{2} \times \frac{2}{3}$

c $3\frac{1}{4} \times 1\frac{2}{3}$

Solution

a $\frac{3 \times 3}{4 \times 5} = \frac{9}{20}$

b $\frac{3 \times 2}{2 \times 3} = \frac{6}{6} = \frac{1}{1} = 1$

c $\frac{13 \times 5}{4 \times 3} = \frac{65}{12} = 5\frac{5}{12}$

Dividing fractions

We have said before that division is the inverse of multiplication.

To divide fractions, we leave the first fraction in the equation alone and turn the division sign into a multiplication sign.

Then we flip the second fraction (the **divisor**) over. This gives its **reciprocal**.

Then multiply the numerators of the two fractions together and then multiply the denominators of the two fractions together.

Finally, simplify the fraction as necessary.

For example, $\frac{1}{2} \div \frac{1}{4}$

$$= \frac{1 \times 4}{2 \times 1} = \frac{4}{2} = 2$$

You can check this by thinking about the pizza. You have $\frac{1}{2}$ left, so how many quarters do you have? Two quarters.

Example

a $\dfrac{3}{4} \div \dfrac{3}{5}$

b $\dfrac{5}{6} \div \dfrac{1}{3}$

c $1\dfrac{1}{4} \div \dfrac{3}{10}$

Solution

a $\dfrac{3 \times 5}{4 \times 3} = \dfrac{15}{12} = 1\dfrac{3}{12} = 1\dfrac{1}{4}$

b $\dfrac{5 \times 3}{6 \times 1} = \dfrac{15}{6} = 2\dfrac{3}{6} = 2\dfrac{1}{2}$

c $\dfrac{5 \times 10}{4 \times 3} = \dfrac{50}{12} = 4\dfrac{2}{12} = 4\dfrac{1}{6}$

Tasks

1 $\quad 2\dfrac{2}{3} + \dfrac{3}{4} =$

2 $\quad \dfrac{2}{7} \times \dfrac{2}{3} =$

3 $\quad \dfrac{7}{10} \div \dfrac{1}{5} =$

4 $\quad 1\dfrac{2}{3} \times \dfrac{4}{5} =$

5 $\quad 2\dfrac{3}{7} + 1\dfrac{2}{5} =$

6 $\quad 1\dfrac{3}{4} + 3\dfrac{5}{8} =$

7 $\quad 5\dfrac{1}{10} - 2\dfrac{3}{10} =$

8 $\quad 4\dfrac{1}{4} - 2\dfrac{3}{8} =$

9 $\quad \dfrac{9}{10} \times \dfrac{3}{4} =$

10 $\quad 2\dfrac{4}{5} \div \dfrac{2}{3} =$

Test your knowledge

1 Solve $2\frac{1}{5} + 2\frac{2}{3}$

  a $4\frac{3}{15}$

  b $4\frac{3}{8}$

  c $4\frac{13}{15}$

  d $5\frac{3}{8}$

2 Solve $\frac{3}{4} \times \frac{1}{2}$

  a $1\frac{1}{2}$

  b $\frac{1}{2}$

  c $\frac{3}{8}$

  d $\frac{1}{4}$

3 Solve $3\frac{5}{6} - \frac{3}{4}$

  a $3\frac{1}{12}$

  b $3\frac{1}{6}$

  c $3\frac{2}{10}$

  d 4

4 Solve $1\frac{4}{5} + 2\frac{3}{10}$

  a $3\frac{7}{15}$

  b $4\frac{1}{5}$

  c $3\frac{1}{10}$

  d $4\frac{1}{10}$

5 Solve $2\frac{2}{3} + 2\frac{1}{5}$

  a $4\frac{3}{15}$

  b $4\frac{3}{8}$

  c $4\frac{13}{15}$

  d $5\frac{3}{8}$

6 Solve $2\frac{1}{3} - 1\frac{6}{7}$

  a $\frac{2}{7}$

  b $\frac{10}{21}$

  c $1\frac{11}{21}$

  d $2\frac{1}{4}$

7 Solve $1\frac{3}{8} \div \frac{1}{4}$

  a $\frac{11}{32}$

  b $2\frac{1}{2}$

  c $4\frac{3}{8}$

  d $5\frac{1}{2}$

8 Solve $2\frac{2}{3} \div \frac{1}{5}$

  a $\frac{8}{15}$

  b $11\frac{1}{8}$

  c $13\frac{1}{3}$

  d $36\frac{2}{3}$

9 What is $\frac{3}{10} \times \frac{4}{5}$?

  a $\frac{3}{8}$

  b $\frac{6}{25}$

  c $1\frac{1}{5}$

  d $\frac{11}{50}$

10 What is $2\frac{1}{2} \times \frac{1}{4}$?

  a $\frac{1}{8}$

  b $\frac{5}{8}$

  c $2\frac{1}{8}$

  d 10

Unit 304
Order of operations and use of calculator

Introduction

Sometimes it is important to know in what order to complete calculations. It does not matter whether you work out 3×5 or 5×3, but $6 \div 3$ gives a different answer than $3 \div 6$ does.

Similarly, the answer to $5 \times 3 - 2$ will be different depending on the order of operations. An operation is what you do with the number such as adding, subtracting, multiplying and dividing or square numbers and square roots.

$5 \times 3 = 15$ and $15 - 2 = 13$

$3 - 2 = 1$ and $5 \times 1 = 5$.

This unit introduces the order of operations: **BODMAS** or **BIDMAS**.

The unit also covers:

- how to use a calculator correctly to get the answer you want
- how to avoid common mistakes using a calculator.

Although you will not be able to use a calculator for the City & Guilds examinations, you will still find calculators useful in everyday life and the examination often has a question based on use of a calculator.

Finally, we look at flow charts as a way of knowing how to complete a task.

Learning objectives

In this unit you will find information on:

- the correct order of operations for calculations, both with and without brackets
- how to use a calculator.

This will help you to prepare for questions involving:

- order of operations for calculations
- reading flow charts.

Ordering mathematical operations

When you work out a sum with more than one operation, you should use BODMAS to ensure you get the same answer as another person completing the sum. BODMAS tells us the order of operations to use:

Brackets first – complete any calculation in brackets before you do any other operations.

Then look at Orders or Indices such as square roots or the power of a number.

Next, complete Division or Multiplication (it does not matter which you complete first). You should always work from left to right.

Next, complete Addition or Subtraction (it does not matter which you complete first). You should always work from left to right.

How do you remember the order? Think BODMAS or BIDMAS.

1	B	Brackets	$(\ldots)$	B	Brackets
2	O	Orders	$\times 2$ $\sqrt{\ }$	I	Indices
3	D	Division	$\div$	D	Division
4	M	Multiplication	$\times$	M	Multiplication
5	A	Addition	$+$	A	Addition
6	S	Subtraction	$-$	S	Subtraction

Example 1

$(3 + 6) \times 5$

Solution

First brackets, so $3 + 6 = 9$. Then $9 \times 5 = 45$

Example 2

$3 + (6 \times 5)$

Solution

First brackets, so $6 \times 5 = 30$. Then $3 + 30 = 33$

Example 3

$3 + 7 \times 4$

Solution

First multiplication, then addition,
so $7 \times 4 = 28$. Then $3 + 28 = 31$

Example 4

$20 - 4^2$

Solution

First orders or indices: 4^2

$4^2 = 16$

So $20 - 16 = 4$

Using a calculator

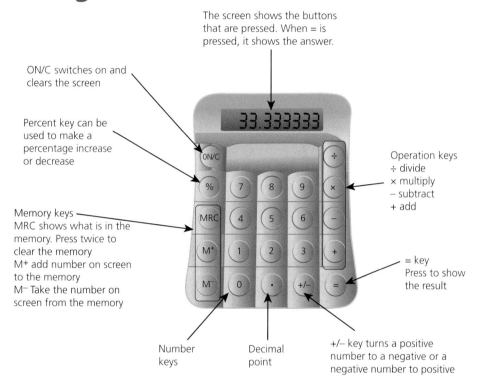

The screen shows the buttons that are pressed. When = is pressed, it shows the answer.

ON/C switches on and clears the screen

Percent key can be used to make a percentage increase or decrease

Memory keys
MRC shows what is in the memory. Press twice to clear the memory
M⁺ add number on screen to the memory
M⁻ Take the number on screen from the memory

Operation keys
÷ divide
× multiply
− subtract
+ add

= key
Press to show the result

Number keys

Decimal point

+/− key turns a positive number to a negative or a negative number to positive

Four simple steps to calculator success:

1 **Estimate**
   Before you touch the calculator, estimate the answer. Round the numbers to find a rough answer.

2 **Clear**
   Switch on and clear the screen. Before you press any buttons, the screen should show a zero '0'.

3 **Calculate**
   Press the buttons on the calculator in the correct order, then the equals button.

4 **Check**
   Look back at your initial estimate. Does the calculator answer look anything like your estimate? If not, press Clear and then calculate again.

Tip for assessment
Remember that you cannot use a calculator in the City & Guilds exams.

Making sense of the answer

The answers for some calculations will be whole numbers. Others may give answers with lots of digits after the decimal point.

For example, put this in a calculator: $25 \div 7 =$

The answer on screen is 3.571428571

There are 9 digits after the decimal point.

The answer is correct to 9 decimal places.

In practice, we do not need to show this level of accuracy most of the time. The level of accuracy depends on the problem.

Some problems need a whole number answer. If we need to work out how many 7-seater taxis are needed for 25 people, the answer 3.571428571 must be rounded up to 4 because 3 of these taxis would only take 21 people.

Dealing with money, we often round to 2 decimal places, the nearest cent.

$\$25 \div 7 = 3.571428571$, which is $3.57 to 2 decimal places.

Activity

Do this calculation on your calculator and on your cellphone.

$7 + 8 \times 4$

Doing calculations in the correct order

The correct order for doing operations is BODMAS or BIDMAS.

If the calculation is $7 + 8 \times 4$ then most calculators will multiply first but some calculators on cellphones do not.

The result using BODMAS (or BIDMAS) is 39. However, on some occasions we may want to add first $7 + 8 = 15$ and then put in $\times 4 =$

In this case, we write the calculation with brackets to make this clear:

$(7 + 8) \times 4$

In BODMAS, we work out brackets first before any other operations. Some calculators have brackets and $(7 + 8) \times 4$ will give the answer 60. Alternatively we need to work out the sum in brackets then multiply this answer by 4.

A number next to the brackets means multiply.

For example, to find the perimeter of a rectangle the formula $P = 2(l + w)$.

Where l is 15 cm and w is 12 cm, what is the perimeter?

We work out the brackets first: press $12 + 15 =$

The answer is 27.

Then multiply by 2, press $\times 2 =$

The answer is 54, so this is 54 cm.

Avoiding common mistakes

It is easy to make mistakes with a calculator. If you press the wrong buttons the answer will be wrong. Try to avoid these common mistakes:

- pressing the wrong number: you want 9 but get 6 on the screen (button below 9).
- pressing a button twice: you want 9 but get 99 on screen.
- missing out a digit: you want 5390, but get 590 on screen.
- putting digits in the wrong order: you want 7410 but get 7401 on screen.
- missing out the decimal point: you want 64.87 but get 6487 on screen.
- pressing the wrong operation key: you want 43×12 but get $43 + 12$ on screen.

Learner tip

Check that a calculation on screen is what you want it to be before pressing =

Estimate to get an idea what you expect the answer to be. If it is different, there may be an error.

Flow charts

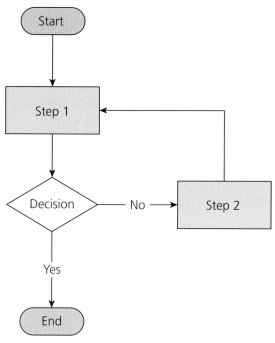

A flow chart always has a start, different steps and at least one decision. Depending on the process, there may be one or more ends or results.

We can put a simple process in the flow chart above to show what to do when you are buying an item in a shop.

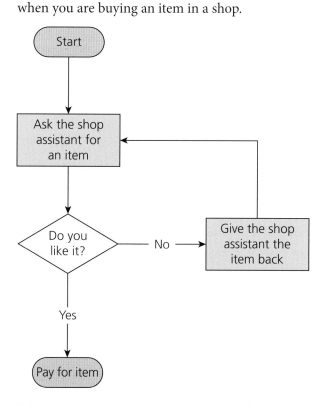

Following the arrows, you leave Start and then Ask the shop assistant for an item.

Now you have to make a decision: Do you like it?

If the answer is Yes, then follow the Yes arrow to pay for the item.

If the answer is No, then follow the No arrow to Give the shop assistant the item back.

Then follow the arrow to Ask the shop assistant for an item.

Now you have to make a decision: Do you like it?

If the answer is Yes, then follow the Yes arrow to pay for item.

If the answer is No, then follow the No arrow to Give the shop assistant the item back.

This takes you back round the flow chart until you find an item you like.

Of course this is very simplistic and you may wish to add more steps, or decisions such as 'Do you have enough money?'

Activity

Design a simple flow chart.

You could ask your partner to check if it works.

Tasks

1 What does BODMAS (or BIDMAS) stand for?

2 You are preparing sandwiches for a party. You decide to have 3 cheese sandwiches and 2 meat sandwiches for each person. There are 30 people. Write the calculation for the number of sandwiches.

3 State three things you should check when using a calculator.

4 Find $(13 + 28) \times 4$

5 Find $(91 - 35) \div 7$

6 Find $14 \times (5 + 16)$

7 Find $(15 + 62) \div (107 - 96)$

8 Create a flow chart to show when and how to round a number.

9 Think of 23.678 rounded to a whole number or 2 decimal places – does your flow chart from question 8 work for this?

10 Discuss your flow chart from question 8 with a partner.

Test your knowledge

1 Find $(28 + 47) \times 3$
   a 78
   b 131
   c 169
   d 225

2 Find $72 \div (18 + 6)$
   a 3
   b 6
   c 10
   d 18

3 Find $28 + (6 \times 19)$
   a 53
   b 142
   c 282
   d 646

4 Find $(15 + 9) \div (10 - 8)$
   a −6
   b 0.3
   c 7
   d 12

5 $2 + 6 \times 4$
   a 22
   b 26
   c 32
   d 36

6 $5^2 + 3$
   a 13
   b 14
   c 28
   d 64

7 2×5^2
   a 20
   b 49
   c 50
   d 100

8 $10 + (12 - 3^2)$
   a 13
   b 16
   c 19
   d 91

9 $10 \div 2 + 3$
   a 2
   b 5
   c 8
   d 23

10 Here are four simple flow charts about lost keys. Which one is correct?

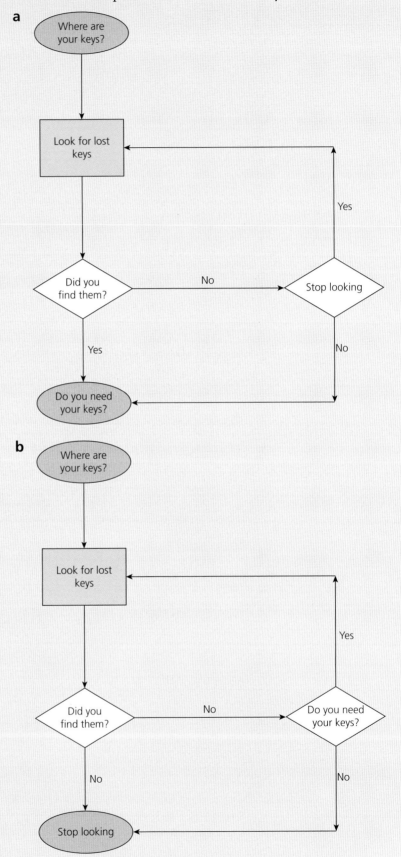

a

b

c

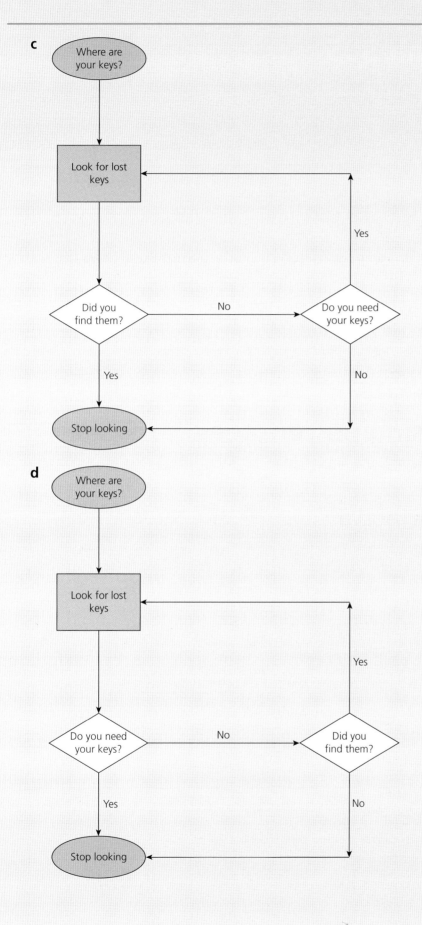

d

Unit 305
Percentages

Introduction

In Unit 206 we looked at expressing numerical information as a percentage and calculating percentages of numbers. You may like to review that Stage 2 unit before starting this unit.

Learning objectives

In this unit you will build on this work and also find information on:

- simple interest
- compound interest
- depreciation.

Using percentages

Try finding 30% of $60

There are different methods you can use.

Method a

1% is $\frac{60}{100} = 0.6$

30% is $30 \times 0.6 = \$18$

Method b

1% is $\frac{60}{100} = 0.6 = 60$ cents

30% is 30×60 cents $= 1800$ cents. $1\,800 \div 100 = \$18$

Method c

Using fractions

$30\% = \frac{30}{100} = \frac{3}{10}$

Then calculate $\frac{3 \times 60}{10 \times 1} = \frac{180}{10} = \$18$

Use whichever method you find easier.

> ## Example
>
> A holiday costs $390 but there is a discount of 6% today.
>
> What does the holiday cost after the discount?
>
> ## Solution (without using a calculator)
>
> 1% is $\frac{390}{100} = 3.9$
>
> 6% is $3.9 \times 6 = 23.4$
>
> 6% is $23.40
>
> The discount is $23.40
>
> So today the holiday costs $390 − $23.40 = $366.60

Interest

There are two types of interest:

- simple interest
- compound interest.

Simple interest is a percentage of the amount invested and is fixed over a period of time. For example, if you deposit $100 in an account paying 2% interest you would earn $2 per year, for each year you invest the money.

Over 3 years, you would earn $3 \times 2 = \$6$

Over 10 years, you would earn $10 \times 2 = \$20$

> ## Example
>
> If you deposit $5000 in an account paying 4% interest per year, how much will you earn over 3 years?
>
> ## Solution
>
> 1% is $50 so 4% is $4 \times 50 = \$200$
>
> Over 3 years, this would be $\$200 \times 3 = \$600$

Banks usually pay **compound interest** to savers. With compound interest, the interest is added to the account each year and so you receive interest on the original deposit plus interest on the interest added each year. The sum from which to calculate interest becomes larger over time.

Example 1

If you deposit $100 in an account paying 2% compound interest, how much interest would you earn over 3 years?

Solution

You would earn $2 in the first year.

In the second year, the interest would be calculated on the new balance in the account,

$100 + $2 = $102

2% interest is $2.04

In the third year, the interest would be calculated on the new balance in the account,

$102 + $2.04 = $104.04

2% interest is $2.0808 which is $2.08

Over 3 years, you would earn $2 + $2.04 + $2.08 = $6.12

Example 2

If you deposit $1 000 in a bank account which is paying 3% compound interest per year. How much interest would you earn over 3 years?

Solution

You can handle this question in either of these ways:

Method a

Calculate the amount of interest for each year and add up all the amounts.

Year one: $1\,000 \times 0.03 = 30$

Year two: $(1\,000 + 30) \times 0.03 = 30.90$

Year three: $(1\,030 + 30.90) \times 0.03 = 31.83$

Total $= 30 + 30.90 + 31.83 = \$92.73$

Method b

Use a multiplier:

Year 3 $= 1\,000 \times 1.03^3 = 1\,092.73$

$1\,092.73 - 1\,000 = \$92.73$

Depreciation

Once a new car leaves the car showroom, it is worth less than the customer paid for it. This is known as **depreciation**. Depreciation is a decrease in the value of something over time. It is usually expressed as a percentage. At Stage 3 we look at simple depreciation, which is also known as **straight line depreciation**. The amount of depreciation is the same each year as it is a percentage of the purchase price, not the current value.

Some items have a longer life than others: machinery, trucks and farming equipment tend to last longer than office equipment and computers, which depreciate more quickly. Accountants calculate a value for a company's assets by estimating depreciation.

Example 1

If an item costing $10 000 usually lasts for 4 years, they may calculate 25% simple depreciation.

Solution

Year	Value at beginning of year $	Depreciation amount 25% $	Value at end of year $
1	10 000	2 500	7 500
2	7 500	2 500	5 000
3	5 000	2 500	2 500
4	2 500	2 500	0

Example 2

Calculate simple depreciation at 10% on an item costing $76 000.

How much will the depreciation be after 2 years?

Solution

Year 1: 10% of 76 000 is 7 600

Year 2: 10% of 76 000 is 7 600

Total depreciation is $7 600 + 7 600 = \$15 200$

An alternative method is to multiply the amount for year 1 by the number of years:

$7 600 \times 2 = \$15 200$

Tasks

1 The cost of making a shirt is $2.60. The manufacturer adds 45% profit and then the retailer adds 20% commission. What is the selling price of the shirt?

2 A pair of shoes cost $64. The shop has a sale with 25% off all their shoes.
   What is the selling price after the discount?

3 Find 6% of $2 420

4 30 mm bolts have a tolerance of 2%. This means the maximum and minimum size of the bolt can be 2% less or more than 30 mm.
   What are the maximum and minimum sizes of the bolts that are still within tolerance?

5 What is 9 out of 20 expressed as a percentage?

6 In a group of 30 people, six do not eat meat.
   What percentage of the people in the group does not eat meat?

7 Calculate simple interest on a $5 000 loan at 8% per annum over 2 years.

8 Calculate compound interest on a $5 000 loan at 8% per annum over 2 years.

9 A machine cost $90 000 when it was new.
   What is the value of the machine after two years simple depreciation at 20% per year?

10 Calculate simple interest on a $2 000 loan at 10% per annum over 3 years.

Test your knowledge

1 A quality control department checks 50 items and finds three have faults.
What is this expressed as a percentage?

   a 3%

   b 5%

   c 6%

   d 47%

2 Calculate compound interest on a $7 000 loan at 5% per annum after 2 years.

   a $350

   b $500

   c $700

   d $717.50

3 A sales person receives $14 000 a year.
He gets a pay rise of 7% next year.
How much is his pay rise?

   a $700

   b $980

   c $1820

   d $2000

4 Find 0.5% of $500.

   a $0.25

   b $2.50

   c $5

   d $50

5 An item costs $240. There is a 20% discount today.
How much does the item cost today?

   a $120

   b $192

   c $200

   d $220

6 What is 6% of $3 000?

   a $180

   b $296

   c $500

   d $1800

7 What is 6 out of 120 expressed as a percentage?

  a 5%

  b 6%

  c 7.2%

  d 20%

8 Calculate the interest on a $5 000 loan for two years at 5% simple interest per year.

  a $250

  b $500

  c $512.50

  d $2 000

9 A machine has a principal cost of $650 000.
Simple depreciation is calculated at 15% per year.
What is the value of two years of simple depreciation?

  a $19 500

  b $300 000

  c $195 000

  d $3 000 000

10 A machine has a principal cost of $200 000.
Simple depreciation is calculated at 20% per year.
What is the value of the machine after two years of simple depreciation?

  a $40 000

  b $120 000

  c $160 000

  d $192 000

Conversions between decimal fractions, common fractions and percentages

Introduction

Fractions, decimals and percentages are used all the time in everyday life. For example, you see shop signs advertising sales of $\frac{1}{2}$ price or 20% off and it is important that you can compare fractions and percentages so that you can work out which is the best offer.

In Unit 207 we looked at recognising **equivalencies** of common fractions, decimal fractions and percentages and how to convert common fractions to decimal fractions.

At Stage 3, we build on this work converting between common fractions, decimal fractions and percentages and use these as appropriate. We also look at some commonly used conversions including recurring decimals.

Learning objectives

In this unit you will find information on:

- converting and using decimal fractions, common fractions and percentages.

Converting between decimal fractions, common fractions and percentages

A fraction such as $\frac{1}{2}$ means a whole divided by two, so a fraction such as $\frac{1}{8}$ means a whole divided by eight. $\frac{3}{8}$ is three of these parts. How can we convert $\frac{3}{8}$ to a decimal fraction?

Thinking about fractions you know will help you to work with other fractions.

$$2\overline{)1.0}^{0.5}$$

> 2 into 1 does not go. Try 2 into 10, there are five 2s in 10.

Remember, the decimal point in the answer goes above the decimal point in the original value.

So the answer is 0.5

Example 1

Now try $\frac{3}{8}$. This is 3 divided by eight.

8 into 3 does not go. Try 8 into 30; there are three 8s in 30 with remainder 6.

Solution

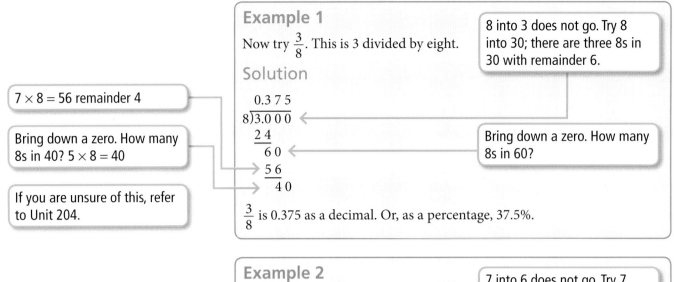

$7 \times 8 = 56$ remainder 4

Bring down a zero. How many 8s in 40? $5 \times 8 = 40$

If you are unsure of this, refer to Unit 204.

Bring down a zero. How many 8s in 60?

$\frac{3}{8}$ is 0.375 as a decimal. Or, as a percentage, 37.5%.

Example 2

What is $\frac{6}{7}$ as a decimal?

7 into 6 does not go. Try 7 into 60; there are eight 7s in 56 with remainder 4.

Solution

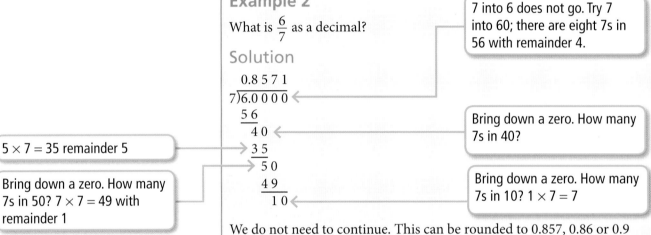

$5 \times 7 = 35$ remainder 5

Bring down a zero. How many 7s in 50? $7 \times 7 = 49$ with remainder 1

Bring down a zero. How many 7s in 40?

Bring down a zero. How many 7s in 10? $1 \times 7 = 7$

We do not need to continue. This can be rounded to 0.857, 0.86 or 0.9 depending on the context and the accuracy needed. Check Unit 208 for more information on rounding if you are unsure about rounding with decimal places.

Learner tip

Remember how to write the division calculation.

$\frac{3}{8}$ is three divided by 8, not 8 divided by 3.

Example 3

Which is the longer distance: half a kilometre or 0.45 kilometres?

Solution

Change both values to fractions or both to decimals. Remember, if you want to compare fractions, they both need to have the same denominator.

$\frac{1}{2}$ is 0.5. Compare 0.5 with 0.45. 0.5 is more than 0.45, so $\frac{1}{2}$ km is further than 0.45 km.

Or, $\frac{1}{2}$ is $\frac{50}{100}$ and 0.45 is $\frac{45}{100}$. $\frac{50}{100}$ is more than $\frac{45}{100}$, so $\frac{1}{2}$ km is further than 0.45 km.

Common conversions including recurring decimals

Sometimes when you divide, the same remainder keeps recurring.

$$
\begin{array}{r}
0.33 \\
3\overline{)1.0} \\
9 \\
\overline{1\,0} \\
9 \\
\overline{1}
\end{array}
$$

> 3 into 1 does not go.
> Try 3 into 10. $3 \times 3 =$
> 9 with one left over.

We can keep dividing by 3 and will always have one left over. This is a **recurring decimal**, also known as a **repeating decimal**. The part that repeats can be shown by placing dots over the first and last digits of the repeating pattern, or by a line over the pattern.

Activity

Find $\frac{2}{3}$ as a decimal number.

Tasks

1 Complete the table.

Decimal	Fraction	Percentage
1		
0.75		
	$\frac{1}{2}$	
		$33\frac{1}{3}\%$
0.25		
0.2		
		10%
	$\frac{2}{3}$	
		5%

2 What is $\frac{5}{8}$ as a decimal fraction?

3 An item costs \$600. Three shops sell the item with different discounts. What is the largest discount?

$\frac{1}{3}$ off   30% off   \$30 off

4 Put these fractions in order of size, smallest first.

0.66 $\frac{2}{3}$ 60%

5 Express 1.65 as a percentage.

Test your knowledge

1 What is 27% expressed as a decimal fraction?

   a 0.027

   b 0.27

   c 2.7

   d 27.00

2 What is $\frac{3}{8}$ expressed as a decimal fraction?

   a 0.375

   b 3.8

   c 0.36

   d 0.03

3 What is 40% as a fraction?

   a $\frac{40}{10}$

   b $\frac{2}{5}$

   c $\frac{4}{5}$

   d $\frac{1}{4}$

4 Which of the following is closest to $\frac{1}{3}$?

   a 1.3

   b 0.13

   c 0.30

   d 0.333

5 What is 0.9 as a percentage?

   a 0.9%

   b 9%

   c 19%

   d 90%

6 What is 0.35 as a fraction?

   a $\frac{3}{5}$

   b 3.5

   c $3\frac{5}{100}$

   d $\frac{7}{20}$

7 A customer wishes to buy an item costing $72.

She has a choice of discount vouchers.

Which discount voucher gives the cheapest price for the item?

a $\frac{1}{3}$ off

b $10 off

c 10% off

d $\frac{1}{4}$ off

8 Which two of the following are equivalent?

A	B	C	D
30%	$\frac{3}{100}$	0.3	$\frac{1}{3}$

a A and B

b B and C

c C and D

d A and C

9 Which two of the following are equivalent?

A	B	C	D
4%	0.4	0.04	$\frac{1}{4}$

a A and B

b A and C

c B and C

d B and D

10 Which two of the following are equivalent?

A	B	C	D
80%	$\frac{1}{8}$	0.08	$\frac{4}{5}$

a A and B

b B and C

c C and D

d A and D

Unit 307
Ratio and proportion

Introduction

This unit builds on Unit 209, which introduced **ratio** and **proportion**, including reading and using scales on maps and plans.

We use proportion in everyday life. Think of sharing a pizza with your friend – do you take one slice each and then another slice each? This is known as direct proportion: one slice for your friend and one slice for you. We can write this as a 1 : 1 ratio.

It would still be direct proportion if you had 2 slices every time your friend had one slice. We could write this 2 : 1.

Scale drawings and scale models use a stated ratio for the drawing. They are used in many different situations where it is important to plan or design something before making it.

- Plans of rooms help design the layout of the furniture.
- Plans or models of buildings show exact details of sizes of walls and windows.
- Diagrams or models of machines show their shape and size compared to other objects.
- A map is a kind of scale drawing that shows where towns and places are located or where buildings and roads are located, depending on the scale of the map.

Learning objectives

This unit introduces inverse proportion and writing a ratio from given data. You will find information on:

- reading and using scales on maps and plans
- using direct proportion and inverse proportion
- writing as a ratio the relationship between two quantities.

This will help you to prepare for questions about:

- calculating distances, foreign exchange rates, ratio and proportions.

Real world maths

Ratios are also used to compare two or more quantities in a variety of situations including baking cakes, mixing cement and diluting concentrated liquids such as cordial or weed killer. In each of these the proportion of one ingredient to the other is stated and happens every time.

Using a scale on maps and plans

A scale drawing is exactly the same shape as the real thing, but smaller. Making a scale drawing is like shrinking an object many times so that it can be shown on an ordinary piece of paper.

The sizes shown on the scale drawing will be in the same proportions as in real life. For example, a window which is half the length of a wall in real life will be half the length of the wall on a scale drawing. This is important, as it means that the layout of the room on the scale drawing will work in real life.

You can also take measurements from a scale drawing and work out the actual measurements from them.

At Stage 2 we looked at simple scales, e.g. 1 centimetre represents 1 metre, which can also be written as a ratio $1:100$

This means that every unit on the drawing is 100 units in real life. We would probably think of this as 1 cm represents 1 metre but this could be any unit such as 1 mm represents 100 mm or 1 inch represents 100 inches.

On maps the scale is much larger because the distances are greater. A typical scale may be $1:25\,000$. This means 1 cm represents $25\,000$ cm. What distance is $25\,000$ cm? It is a quarter of a kilometre.

Other common scales are $1:50\,000$ and $1:75\,000$.

Example 1

The scale on the plan is $1:50$

A wall measures 3 m in real life. How big should this be drawn on the plan?

Solution

$1:50$ means 1 unit represents 50 units. We can say 1 cm represents 50 cm (or $\frac{1}{2}$ m).

The measurement of the wall is in metres but the measurement on the plan is likely to be centimetres. Remember there are 100 cm in a metre.

If 1 cm represents 50 cm $\left(\frac{1}{2}\,\text{m}\right)$

then 2 cm represents 100 cm (or 1 m)

3 cm represents 150 cm $\left(\text{or } 1\frac{1}{2}\,\text{m}\right)$

4 cm represents 200 cm (or 2 m)

5 cm represents 250 cm $\left(\text{or } 2\frac{1}{2}\,\text{m}\right)$

6 cm represents 300 cm (or 3 m)

Or, we could divide 3 m by 50 cm – make sure the units are the same before you divide.

$300 \div 50 = 6$, so the window should be drawn as 6 cm on the plan.

Example 2

On a map of scale 1 to 75 000, the distance between two towns is represented by a line 6 cm long. What is the actual distance?

Solution

If 1 represents 75 000 then 1×6 represents $75\,000 \times 6$

So 6 cm represents $450\,000 = 4.5$ km

Using direct proportion

With direct proportion, as one amount increases the other amount increases at the same rate.

Ratios are usually written as simply as possible.

If you have one sweet every time your friend has one sweet, this ratio is $1:1$

If there are six sweets and you share them in a $1:1$ ratio, you would have 3 each. $3:3$

We could work this out:

$1:1$ (2 sweets in total)

$2:2$ (4 sweets in total)

$3:3$ (6 sweets in total)

Or we could divide the total number of sweets (6) by the number each time we share them out (2)

$6 \div 2 = 3$

We must do the same to each side of the ratio so

$1 \times 3:1 \times 3$

3 sweets : 3 sweets

We know from this that if you had 10 sweets and shared them in a $1:1$ ratio, you and your friend would both have 5 sweets.

Example 1

In a recipe, the ratio of weight of flour to weight of sugar is $2:1$

If I use 400 g flour, how much sugar do I need?

Solution

This time you know the increase on one of the sides of the ratio, the flour.

$400 \div 2 = 200$ so

$2 \times 200:1 \times 200$

$400:200$

400 g flour : 200 g sugar

Example 2

In a recipe, the ratio of weight of flour in grams to number of eggs is 300 : 2. If I use 6 eggs, how much flour do I need?

Solution

$6 \div 2 = 3$ so

$3 \times 300 : 2 \times 3$

$900 : 6$

900g flour : 6 eggs

We can also work out how much an amount of money is worth in a different currency if we know the exchange rate.

Example 3

1 US dollar is approximately 0.89 euro.

How many euros would I receive for 100 US dollars?

Solution

$100 \div 1 = 100$

Multiply each side by the same number.

$1 \times \$100 = €0.89 \times 100$

$\$100 = €89$

Using inverse proportion

Inverse proportion is where one amount decreases at the same rate that the other increases.

If you employ more workers, the job should be completed sooner. Therefore, the number of workers to the time taken is inversely proportional.

Example 1

If it takes 2 hours for 4 workers to complete a job, how long will it take 8 workers to complete the job?

Solution

Use the information you know:

$$\text{Time} = \frac{x}{\text{number of people}}$$

$$2 \text{ hours} = \frac{x}{4 \text{ people}}$$

Rearrange the equation to get x (the number of work hours) by itself:

$4 \times 2 = x$

$x = 8$, so we can substitute this in the new equation:

$$\text{Time in hours} = \frac{x}{\text{number of people}}$$

$$\text{Time} = \frac{8}{8}$$

Time = 1 hour. It will take 1 hour for 8 workers to complete the job.

Example 2

If it takes 3 hours for 4 workers to complete a job, how long will it take 6 workers to complete the job?

Solution

Use the information you know

$$\text{Time} = \frac{x}{\text{number of people}}$$

$$3 \text{ hours} = \frac{x}{4 \text{ people}}$$

Rearrange the equation to get x (the number of work hours) by itself:

$$3 \times 4 = x$$

$x = 12$, so we can substitute this in the new equation:

$$\text{Time in hours} = \frac{x}{\text{number of people}}$$

$$\text{Time} = \frac{12}{6}$$

Time = 2 hours. It will take 2 hours for 6 workers to complete the job.

Speed and travel time are inversely proportional because, for example, the faster a car goes the shorter the time it takes to travel a certain distance. As speed goes up, travel time goes down and as speed goes down, travel time goes up.

Example 3

A journey takes 4 hours at 30 km per hour.

How long does it take at 50 km per hour?

Solution

$$\text{speed} = \frac{\text{distance}}{\text{time}}$$

$$30 = \frac{\text{distance}}{4}$$

$$30 \times 4 = \text{distance}$$

$$120 = \text{distance}$$

Substitute the value for distance in the new equation

$$50 = \frac{120}{\text{time}}$$

$$50 \times \text{time} = 120$$

$$\text{time} = \frac{120}{50}$$

time = 2.4 hours or 2 hours 24 minutes

Simplifying ratios

In your exam, you may be asked to express some information as a ratio in its simplest form. This means you should try to divide the numbers until you are left with the number one, or the lowest number you can get, on one side.

Example 1

A catering assistant uses 3 spoons of lemon juice to 12 spoons of water. What is this ratio in its simplest form?

Solution

The ratio of lemon juice to water is $3:12$

Both numbers divide by 3 so

$3 \div 3 : 12 \div 3$

$1:4$

The ratio of lemon juice to water is $1:4$ in its simplest form.

Example 2

Here's a different ratio. A catering assistant uses 10 spoons of lemon juice to 15 spoons of water. What would this ratio be in its simplest form?

Solution

The ratio of lemon juice to water is $10:15$

Both numbers divide by 5 so

$10 \div 5 : 15 \div 5$

$2:3$

The ratio of lemon juice to water is $2:3$ in its simplest form.

Example 3

Sometimes the numbers are much larger, e.g. 36 parts to 96 parts.

What would this ratio be in its simplest form?

Solution

$36:96$

Do you know what number each of these values can be divided by with no remainders?

If not, try 2

$18:48$ then 2 again

$9:24$

We cannot divide 9 by 2 so try 3

$3:8$

You could have divided 36 and 96 by 12, if you had recognised they were both divisible by 12. The answer is the same.

Activity

Find the lowest ratio of $36:96$ by dividing by 3 first.

Tasks

1 A room is drawn to a scale of 1:50
A wall of the room is 8 centimetres long on the scale drawing.
How long is the actual wall?

2 A map has a scale of 1:200 000.
The distance between two towns on the map is 5.2 cm.
What is the actual distance between the two towns?

3 A map has a scale of 1:250 000.
On the map, the distance between two places is 2.6 cm.
What is the actual distance?

4 A model of a ship has a scale of 1:80.
The model is 75 cm long.
What is the actual length of the ship?

5 You have 24 coins. You agree that for every one coin that you give your brother you will keep three.

a What is the ratio?

b How many coins will you have?

c How many coins will your brother have?

6 Write this ratio in its simplest form:
6:12:8

7 A journey takes 3 hours at 30 km per hour.
How long does it take at 40 km per hour?

8 Grace wants to draw a plan of her bathroom floor.
The bathroom floor is a rectangle 3 m by 4 m.
She uses a scale of 1:20.
What will the size of the rectangle be on the plan?

9 The exchange rate is 1 US dollar = 0.88 euro.
How many euros will I receive for 400 US dollars?

10 Write this ratio in its simplest form:
36:60

Test your knowledge

1 A plan of a room has a scale of 1 : 40.
   One wall has a length of 16 cm on the plan.
   What is the actual length of the room?

   a 1.6 m

   b 2.5 m

   c 3.2 m

   d 6.4 m

2 A drawing of an office has a scale of 1 : 50
   The actual office has a length of 10 m.
   What is the length of the office on the drawing?

   a 2 cm

   b 5 cm

   c 20 cm

   d 50 cm

3 A model of a new airport has a scale of 1 : 5000.
   On the model, a runway has a length of 80 cm.
   What is the actual length of the runway?

   a 1.6 km

   b 4 km

   c 6.25 km

   d 16 km

4 A room has a length of 6.2 metres and width 4.7 metres.
   A scale plan of the room will be made on paper of length 40 cm and
   width 28 cm.
   Which of these scales gives the largest plan that will fit on the paper?

   a 1 : 10

   b 1 : 20

   c 1 : 25

   d 1 : 40

5 A cook uses 2 eggs to make 24 small cakes.
   How many cakes will 5 eggs make?

   a 6

   b 10

   c 12

   d 60

6 An orange drink is mixed in the ratio of 25 ml of orange cordial to 1 litre of water.

What is the ratio of orange cordial to water in its simplest form?

a 25:1

b 1:25

c 4:1

d 1:40

7 It takes 8 hours for 4 workers to complete a job.

How long will it take 6 workers to complete the job?

a 3 hours

b 5 hours 12 minutes

c 5 hours 20 minutes

d 12 hours

8 A builder mixes cement with sand using a 1:6 ratio.

He uses three shovels of cement.

How many shovels of sand does he need?

a 2

b 3

c 8

d 18

9 On a map with a scale of 1:75 000 the distance between a hotel and the airport is 3 cm.

What is the actual distance in km?

a 2.25 km

b 2.5 km

c 22.5 km

d 25 km

10 A journey takes 2.5 hours at an average speed of 50 km/h.

How long will the journey take at 30 km/h?

a About 2 hours

b Just over 3 hours

c Just over 4 hours

d About 6 hours

Unit 308
Measurement and standard units

Introduction

At Stage 2, we used measure and time to solve everyday problems in Unit 202 and found the area and perimeter of L-shaped rooms and the volume of cuboids in Unit 212.

Stage 3 builds on this work using metric and imperial measures of

- length
- area and perimeter
- volume
- weight or mass
- capacity
- time
- temperature.

Learning objectives

In this unit, you will find information on:

- the area of triangles
- the area of circles
- the volume of cylinders
- time in different time zones

This will help you to prepare for questions about:

- solving problems related to length, area, volume, weight, capacity or temperature in metric or imperial units
- working out time differences between countries.

You may find it useful to look through Unit 202 and Unit 212 before you start working on this unit.

Length

At Stage 3, you should be familiar with these metric measurements: kilometres, metres, centimetres and millimetres.

Fill in the gaps in this table.

mm	cm	m	km
			1
		1	
	10		
		100	
	1		
1			

Check your answers with the table at the end of this unit.

You should also be familiar with these imperial measurements:

12 inches = 1 foot
3 feet = 1 yard
1760 yards = 1 mile

At Stage 3 you will be given conversion rates for imperial units if you need to work with them.

Example 1

A wall measures 10 m wide with a height of 2.4 m.

2.4 m

10 m

A roll of wallpaper is 10.05 m long and 52 cm wide.

The wallpaper needs to be hung vertically from top to bottom of the wall with no joins in the length of each sheet.

How many rolls of wallpaper are needed?

Solution

Because you cannot join a length of wallpaper part way down the wall, we need to calculate how many lengths of wallpaper can be cut from one roll rather than dividing the area of the wall by the area covered by one roll of wallpaper.

$10.05 \div 2.4 = 4$ whole lengths or drops of wallpaper in one roll of wallpaper.

Now find how many widths of wallpaper are needed for the whole wall.

$10 \text{ m} \div 52 \text{ cm} = 1000 \div 52 \text{ cm} = 19.6$

So 20 lengths or drops are needed.

There are 4 whole lengths or drops of wallpaper on a roll.

So $20 \div 4 = 5$ rolls of wallpaper are needed.

Example 2

A fence panel measures 6 ft × 6 ft. One fence panel will be used to make a gate.

How many fence panels are needed to go around this back yard?

Solution

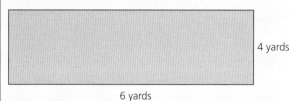

6 yards

4 yards

The perimeter of the back yard is 6 + 4 + 6 + 4 = 20 yards

One fence panel is 6 feet, which equals 2 yards. Divide the total perimeter by 2.

10 fence panels are required.

Area

At Stage 2 you calculated the area of squares, rectangles and composite shapes. Now we will calculate the area of triangles and circles.

Area of a triangle

The area of a square or rectangle is length × width.

What is the area of a triangle?

Example 1

Find the area of this triangle.

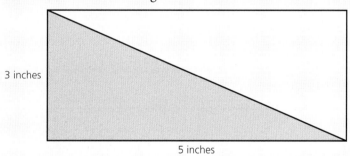

3 inches

5 inches

Solution

This triangle is half a rectangle, so we can find the area of the rectangle and divide by two.

Area of a triangle = $\frac{1}{2}$ base × height

3 × 5 = 15 square inches

15 ÷ 2 = 7.5 square inches

Area of a circle

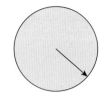

diameter   radius

The distance from one side of a circle to the other is called the **diameter**.

The distance from the middle of a circle to the side is called the **radius**.

The area of a circle is πr^2

This means $\pi \times r \times r$

π is usually rounded to 2 or 3 decimal places. At Stage 3, you can use π as 3.14 or $\frac{22}{7}$.

Example 2

In this example, the radius r is 6 cm. Find the area.

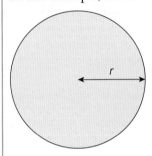

Solution

The area of a circle is πr^2

Note: using BODMAS we usually find r^2 first.

$r^2 = 6^2 = 36$

$\pi r^2 = \pi \times 36$

$A = 113.04\,\text{cm}^2$

Volume

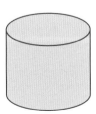

At Stage 2 we said that the volume of a cube or cuboid = length × width × height

which is the same as the area of the base × height.

This 3D shape is a cylinder.

The volume of a cylinder is the area of the base × height. The area of the base of a cylinder is the area of a circle.

> ## Example
>
> The radius of a cylinder is 5 cm and the height is 10 cm.
>
> What is the volume of the cylinder?
>
> ## Solution
>
> The volume of a cylinder is $\pi r^2 h$
>
> $= 3.14 \times 5 \times 5 \times 10$
>
> $= 785 \text{ cm}^3$

Weight

The metric units for weight are grams and kilograms. For larger weights there are tonnes.

There are:

1000 grams (g) in 1 kilogram (kg)

1000 kilograms (kg) in 1 tonne.

The imperial units are ounces, pounds, stones and tons.

There are:

16 ounces (oz) in 1 pound (lb)

14 pounds (lb) in 1 stone

2240 pounds (lb) in 1 ton.

Stones are generally only used for people's weight.

1 metric tonne is 0.984 207 imperial tons.

At Stage 3 you will be given conversion rates for imperial units if you need to work with them.

> ## Example
>
> A cook puts the following ingredients into a bowl.
>
> 6 oz butter
>
> 6 oz lard
>
> 12 oz flour
>
> What is the total weight of the ingredients?
>
> ## Solution
>
> There are 16 ounces (oz) in 1 pound (lb).
>
> $6 + 6 + 12 = 24 \text{ oz}$
>
> $\qquad\quad = 1 \text{ lb } 8 \text{ oz}$
>
> $\qquad\quad = 1\frac{1}{2} \text{ lb}$

Capacity

Volume is the space taken up by the object itself, while capacity refers to the amount of substance, like a liquid or a gas, which a container can hold. A jug may have a capacity of 2 litres (it is capable of holding 2 litres of water) but only 1 litre of water is in the jug.

We could say the volume of liquid in the jug is 1 litre but volume is often measured in cubic units, while capacity can be measured in almost every other unit, including litres, gallons, pounds, etc.

The metric units of capacity are litres (l), centilitres (cl) and millilitres (ml).

There are:

100 cl in a litre and

1000 ml in a litre.

The imperial units for capacity are pints and gallons.

There are:

8 pints in a gallon.

At Stage 3 you will be given conversion rates for imperial units if you need to work with them.

Example

A pharmacy assistant dispenses cough medicine from a 2-litre container into 100 ml bottles.

There are 1000 ml in a litre.

a How many 100 ml bottles can she fill?

b How many 5 ml spoonfuls of medicine will be in each bottle?

Solution

a 2 litres is 2000 ml

$2000 \div 100 = 20$

Each container can fill 20 bottles.

b Each bottle is 100 ml, so

$100 \div 5 = 20$

Each bottle contains twenty 5 ml spoonfuls.

Temperature

We measure temperature in degrees Celsius and degrees Fahrenheit.

You may be required to read a temperature below 0 °C from a thermometer.

Example 1

What temperature does this thermometer show?

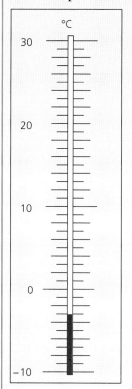

Solution

A reading below 0 °C indicates a negative temperature.

On this thermometer, there are 10 small unlabelled divisions for every 10 degrees, so each small division is 1 °C.

The level of the liquid is 3 divisions below 0, so the temperature is −3 °C.

For more on working with negative numbers see Unit 1 for the section on directed numbers.

You may be given the formula to convert temperatures from one system to the other:

Fahrenheit = 1.8 °C + 32.

Example 2

The washing instructions on an item of clothing say 'wash at 30 °C'.

Approximately what is this temperature in °F?

(Fahrenheit = 1.8 °C + 32)

Solution

$1.8 \times 30 = 54$

$54 + 32 = 86$ °F

Time

At Stage 2 you used this information to help you solve time problems.

60 seconds = 1 minute

60 minutes = 1 hour

24 hours = 1 day

7 days = 1 week

365 days = 1 year (or 366 days in a leap year)

12 months = 1 year

At Stage 3 time problems include using different **time zones**. In different parts of the world, the time is different. When it is 8:00 a.m. in London, it is 9.00 a.m. in Paris, 3.00 a.m. in New York and 5.00 p.m. in Tokyo.

The world is divided into time zones. A time zone is a region where the same standard time is used. Time zones are often based on boundaries of countries or lines of longitude.

When you move from one time zone to another, the time will be different.

This map shows approximate time zones.

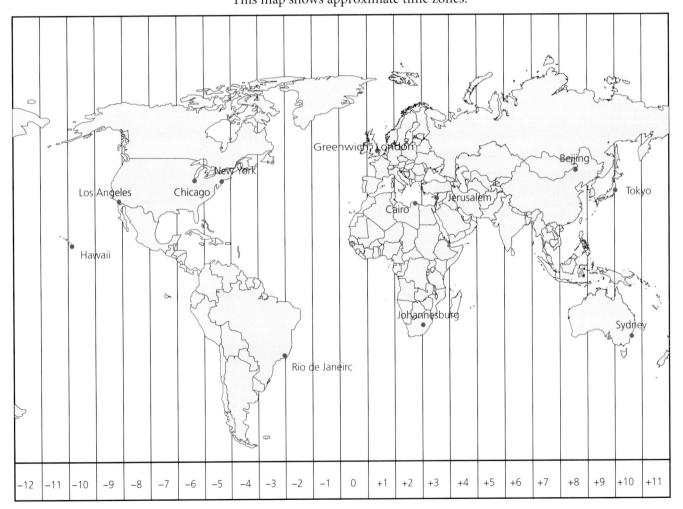

The times are based on Greenwich in London. This is because times around the world were first calculated in England.

The UK is on Greenwich Mean Time (GMT). The time zone is labelled 0 on the map. Other time zones are labelled with a number.

Time zones to the **West** of London, UK have **negative** numbers.

This means that the time is **earlier** than in the UK.

> ## Example 1
>
> New York is in the zone labelled −5.
>
> What time is it in New York when it is 8.00 p.m. in London?
>
> What time is it in New York when it is 3.00 a.m. in London?
>
> ## Solution
>
> The time in New York is 5 hours earlier than in London.
>
> When it is 8.00 p.m. in London, it is 3.00 p.m. in New York.
>
> At 3.00 a.m. today in London, it is 10.00 p.m. yesterday in New York.

Time zones to the East of London have positive numbers.

This means that the time is later than in London.

▲ Beijing

> ## Example 2
>
> Beijing is in the zone labelled +8. What time is it in Beijing when it is 2 p.m. in London?
>
> What time is it in Beijing when it is 10 p.m. in London?
>
> ## Solution
>
> The time in Beijing is 8 hours later than London.
>
> At 2.00 p.m. in London, it is 10.00 p.m. in Beijing.
>
> At 10.00 p.m. today in London, it is 6.00 a.m. tomorrow in Beijing.

Planning journeys and events, and calculating schedules for work or leisure activities, can involve complicated time problems.

Remember: ←

- the times may be expressed in different ways, 12-hour or 24-clock, hours or hours and minutes or fractional hours
- they may involve using information on time zones.

> It is important to work:
>
> - step by step to avoid forgetting any part or making errors
> - in common units – either minutes or hours and minutes
> - in common time zone – dealing with any time differences.

Example 3

A student is travelling around the world. The next stage of her journey is travelling from Hull in the UK to Holland.

A ferry leaves Hull in the UK at 8.30 p.m. and arrives in Rotterdam at 8.15 a.m. the next day, local time.

Time in Holland is 1 hour later than in the UK.

How long does the ferry crossing last?

Solution

To find the answer, we need to deal with the time difference.

The arrival time is 8.15 a.m. local time.

This time is one hour later than in the UK.

In UK time, the arrival is at 7.15 a.m.

Next, find the difference between the departure time and the arrival time.

A simple method is counting up from the departure time.

8.30 p.m. to 9.00 p.m. is 30 minutes.

9.00 p.m. to 12.00 a.m. is 3 hours.

12.00 a.m. to 7.15 a.m. is 7 hours 15 minutes.

30 minutes + 3 hours + 7 hours 15 minutes = 10 hours 45 minutes

The ferry crossing lasts 10 hours 45 minutes or $10\frac{3}{4}$ hours

Tasks

1 You need to buy a new carpet for an L-shaped room.

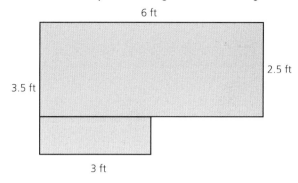

6 ft

2.5 ft

3.5 ft

3 ft

   a What is the total area of the room?

   b What is the perimeter of the room?

2 What is the circumference of this circle?

(Assume $\pi = 3.14$)

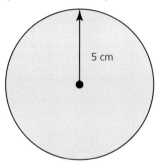

5 cm

3 What is the volume of this chest?

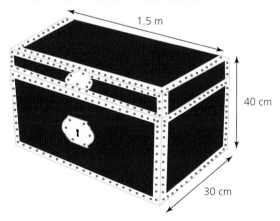

1.5 m

40 cm

30 cm

4 An office has a small area of land outside measuring 4.5 m by 6 m. The weeds on the land need to be killed before it can become a mini picnic area.

A tin of weed killer kills 9 square metres (m²) and costs $4.

How much will it cost to kill off all the weeds?

5 What is the volume of the cylinder below?

(Assume π = 3.14)

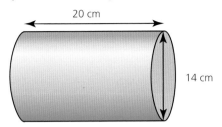

20 cm

14 cm

6 This thermometer shows the temperature in a warehouse.

Yesterday, the temperature was 6 °C.

What is the change in temperature since yesterday?

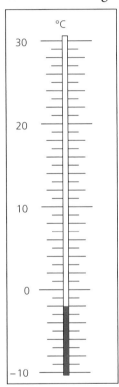

7 A group of volunteers runs an advice centre.

The centre is open from 9.00 a.m. to 2.00 p.m. every day from Monday to Saturday.

When the centre is open, there must be one volunteer between 9.00 a.m. and 12.00 p.m. and two volunteers between 12.00 p.m. and 2.00 p.m.

This table shows when two volunteers, Kara and Tim, will be in the centre next week.

Volunteer	Kara	Tim	Jamie
Monday	09:00 to 14:00		
Tuesday	09:00 to 12:00	12:00 to 14:00	
Wednesday	12:00 to 14:00	12:00 to 14:00	
Thursday	12:00 to 14:00	09:00 to 14:00	
Friday		12:00 to 14:00	
Saturday		12:00 to 14:00	
Total hours			

The manager, Jamie, will work at the centre at the times when there are not enough volunteers.

a Work out what times Jamie needs to be at the centre.

b What is the total number of hours Jamie needs to be at the centre?

8 What is the total area of this playground?

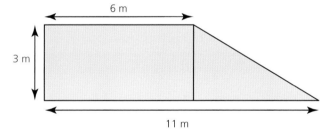

9 The time in Kingston, Jamaica, is one hour behind New York, USA. When it is 11.30 p.m. in New York, what time is it in Kingston?

10 How many 300 g packs of birdseed can a shopkeeper make from 10 kg? How much is left over?

1 What is the area of a circle with a diameter of 12 cm?
(Assume $\pi = 3.14$)

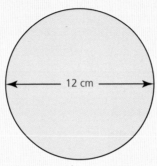

12 cm

a 37.68 cm²		**c** 113.04 cm²	
b 59.16 cm²		**d** 452.16 cm²	

2 Find the area of this triangle.

4 cm

8 cm

a 12 cm²		**c** 24 cm²	
b 16 cm²		**d** 32 cm²	

3 What is the volume of this cylinder?
(Assume $\pi = 3.14$)

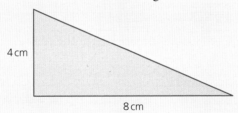

0.3 m

20 cm

a 94.2 cm³		**c** 9 420 cm³	
b 376.8 cm³		**d** 37 680 cm³	

4 A shop assistant has 5 yards of ribbon. She uses 12 inches to make one bow.
What is the maximum number of bows she can make?
(12 inches = 1 foot, 3 feet = 1 yard)

a 5		**c** 20	
b 15		**d** 60	

5 A bottle of water contains 75 cl.
How much water is there in 15 of these bottles?
Give your answer to the nearest litre.

  a 20 litres   **c** 5 litres

  b 11 litres   **d** 1 litre

6 The washing instructions on an item of clothing say 'wash at 40 °C'.
Approximately what is this temperature in °F?
(Fahrenheit = 1.8C + 32)

  a 40 °F   **c** 72 °F

  b 50 °F   **d** 104 °F

7 What is the circumference of a circular pond with a diameter of 12 m?
(Assume π = 3.14)

  a 3.82 m   **c** 48.0 m

  b 37.68 m   **d** 113.04 m

8 A cook uses 200 ml of milk to make one portion of pudding.
He has a 2-litre container of milk.
How many portions can he make?

  a 4   **c** 100

  b 10   **d** 400

9 The time in Manchester UK is 5 hours ahead of Kingston, Jamaica.
If the time in Jamaica is 11.39 a.m. on Monday, what is the time in the UK?

  a 6.39 a.m. Monday   **c** 4.39 a.m. Tuesday

  b 4.39 p.m. Monday   **d** 6.39 p.m. Monday

10 A film club plans a special film double bill evening. They will show two films with a break of $\frac{1}{4}$ hour between them.
The first film last for 101 minutes.
The second film lasts for 139 minutes.
They must end before 11.00 p.m.
What is the latest time they can start the first film?

  a 6.00 p.m.   **c** 7.15 p.m.

  b 6.45 p.m.   **d** 8.05 p.m.

Unit 309
Reading and interpreting tables of figures, dials and scales

Introduction

This unit builds on the work on tables, dials and scales from Stage 2 including Unit 213.

Learning objectives

In this unit, you will find information on:

- extracting information from a variety of tables including timetables, conversion tables, town/distance tables, information tables
- reading dials and scales on a variety of measuring devices and meters.

Tables

Timetables

Managing your time is a very important skill, so it is common to see data handling questions which involve time. It is usually possible to use a number of different strategies to solve the tasks in exams, including trial and error, and there will often be more than one correct solution.

Read the headings across the top and down the side of the timetable carefully and you should be able to answer the questions.

Example 1

	Mon 2 October	Tues 3 October	Wed 4 October	Thurs 5 October	Fri 6 October	Sat 7 October	Sun 8 October
Latitia	kitchen 12:00 p.m.–6:00 p.m	kitchen 12:00 p.m.–6:00 p.m	kitchen 12:00 p.m.–6:00 p.m	kitchen 12:00 p.m.–6:00 p.m	Rostered day off	on call	
Tedroy	kitchen 8:00 a.m.–1:00 p.m	kitchen 8:00 a.m.–1:00 p.m	kitchen 8:00 a.m.–1:00 p.m	kitchen 8:00 a.m.–1:00 p.m	kitchen 8:00 a.m.–1:00 p.m		
James	on call	on call			on call	kitchen 9.00 a.m.–2.00 p.m	kitchen 10.00 a.m.–3.00 p.m
Matthew	Bar 8:00 a.m.–1:00 p.m	Bar 8:00 a.m.–1:00 p.m	Bar 8:00 a.m.–1:00 p.m	on call	Bar 12:00 p.m.–6:00 p.m	kitchen 1:00 p.m.–6:00 p.m	kitchen 1:00 p.m.–5:00 p.m
Shelley	Bar 9:00 a.m.–6:00 p.m 1 hour break	Bar 9:00 a.m.–6:00 p.m 1 hour break	Bar 9:00 a.m.–6:00 p.m 1 hour break	Bar 9:00 a.m.–6:00 p.m 1 hour break	kitchen 12:00 p.m.–6:00 p.m	on call	
Charlie	Ski Holiday	Ski Holiday	Ski Holiday	Ski Holiday	Ski Holiday		
Charlene	on call	on call	on call	Bar 8:00 a.m.–1:00 p.m	Bar 8:00 a.m.–1:00 p.m	Weekend Off	Weekend Off
Leon	kitchen 9:00 a.m.–6:00 p.m 1 hour break	kitchen 9:00 a.m.–6:00 p.m 1 hour break	kitchen 9:00 a.m.–6:00 p.m 1 hour break	Sick	kitchen 9:00 a.m.–6:00 p.m 1 hour break		on call
Sarah					Training 10:00 a.m.–4:00 p.m	Training 9:00 a.m.–2:00 p.m	Bar 10:00 a.m.–3:00 p.m
Nick	Training 10:00 a.m.–4:00 p.m	Training 10:00 a.m.–4:00 p.m	Bar 10:00 a.m.–4:00 p.m	Bar 10:00 a.m.–4:00 p.m	Bar 10:00 a.m.–4:00 p.m	Bar 01:00 p.m.–6:00 p.m	Bar 01:00 p.m.–5:00 p.m

a What time is Matthew working on 6 October?
b If Latitia can't attend for her shift on 4 October, who is on call?

Solution

a Follow a line across from Matthew and another down from Friday 6 October. Where the two lines meet is the time Matthew is working that day. He is working from 12.00 p.m–6.00 p.m.

b Look down the column for 4 October until you find the 'on call' box. Follow this line back to the names. Charlene is on call that day.

Example 2

What is 1 mile in metres?

METRIC CONVERSIONS						
1 centimetre	=	10 millimetres		1 cm	=	10 mm
1 decimetre	=	10 centimetres		1 dm	=	10 cm
1 metre	=	100 centimetres		1 m	=	100 cm
1 kilometre	=	1000 metres		1 km	=	·1000 m

IMPERIAL CONVERSIONS						
1 foot	=	12 inches		1 ft	=	12 in
1 yard	=	3 feet		1 yd	=	3 ft
1 chain	=	22 yards		1 ch	=	22 yd
1 furlong	=	220 yards (or 10 chains)		1 fur	=	220 yd (or 10 ch)
1 mile	=	1760 yards (or 8 furlongs)		1 ml	=	1760 yd (or 8 fur)

METRIC ⟶ IMPERIAL CONVERSIONS						
1 millimetre	=	0.03937 inches		1 mm	=	0.03937 in
1 centimetre	=	0.39370 inches		1 cm	=	0.39370 in
1 metre	=	39.37008 inches		1 m	=	39.37008 in
1 metre	=	3.28084 feet		1 m	=	3.28084 ft
1 metre	=	1.09361 yards		1 m	=	1.09361 yd
1 kilometre	=	1093.6133 yards		1 km	=	1093.6133 yd
1 kilometre	=	0.62137 miles		1 km	=	0.62137 mi

IMPERIAL ⟶ METRIC CONVERSIONS						
1 inch	=	2.54 centimetres		1 in	=	2.54 cm
1 foot	=	30.48 centimetres		1 ft	=	30.48 cm
1 yard	=	91.44 centimetres		1 yd	=	91.44 cm
1 yard	=	0.9144 metres		1 yd	=	0.9144 m
1 mile	=	1609.344 metres		1 mi	=	1609.344 m
1 mile	=	1.609344 kilometres		1 mi	=	1.609344 km

Solution

This is imperial to metric so use the second half of the table.

There are two options for 1 mile. Make sure you choose the correct option for metres.

1 mile = 1609.344 metres

Example 3

This is part of a flight timetable for flights from Kingston (KIN), Jamaica to JFK Airport, New York, USA.

03:05 – 06:49	JetBlue 960	M	T	W	T	F	S	S	KIN – JFK
05:30 – 09:30	Fly Jamaica Airways 270	M	–	–	–	–	–	–	KIN – JFK
06:00 – 10:00	Fly Jamaica Airways 272	–	–	–	T	–	S	S	KIN – JFK
10:30 – 14:15	JetBlue 560	M	T	W	T	F	S	S	KIN – JFK
15:36 – 19:25	JetBlue 2960	M	–	–	T	F	S	S	KIN – JFK
19:25 – 23:10	Caribbean Airlines 17	M	T	W	T	F	S	S	KIN – JFK

I want to leave Kingston on Friday morning. What time planes can I catch?

Solution

Look at the day columns for F. When there is a no F there, it means that plane does not fly at that time on a Friday.

There are only two planes leaving on Friday morning: 03:05 and 10:30.

Example 4

This grid shows the distances between towns in miles.

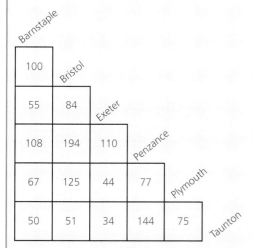

A business man is flying to Bristol airport. How far is it from Bristol to his meeting in Plymouth?

Solution

Draw a line down from Bristol and another across from Plymouth. Where the two lines meet is the distance between the two towns.

From Bristol to Plymouth is 125 miles.

Dials and scales

At Stage 3, you will be expected to read a variety of familiar and unfamiliar scales. In each case, you should think about what each small division represents on the scales.

Using weighing instruments often means reading a scale.

This diagram shows the sort of scale found on weighing machines.

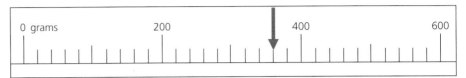

The pointer shows the weight of the item.

The pointer is between 200 grams and 400 grams.

To work out the exact weight, we need to know what weight the divisions show (shown by the small lines).

There are 10 divisions between 200 grams and 400 grams.

Those 10 divisions cover 200 grams.

So, each division is 20 grams.

You can check by counting 20s for each division. The first division after 200 grams is 220 grams, then the next is 240 grams and so on up to 400.

There are 8 divisions between 200 grams and the pointer.

8 divisions are 160 grams.

160 grams more than 200 is 360 grams. That is the weight shown.

To check, there are 2 divisions between the pointer and 400 grams.

2 divisions are 40 grams.

40 grams less than 400 grams is 360 grams.

On different weighing machines, the unlabelled divisions might not be 20 grams. They could be 5 grams, 10 grams, 50 grams, 250 grams, 0.1 kg, 2 kg etc.

Example 1

What is each small division on these scales?

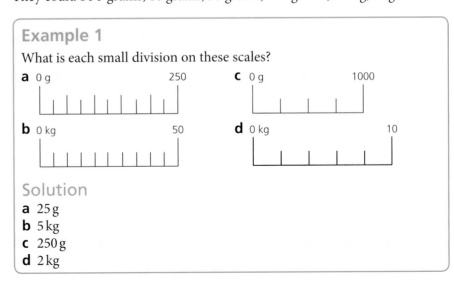

Solution
a 25 g
b 5 kg
c 250 g
d 2 kg

Example 2

Marva weighs herself.

The pointer on this scale shows her weight.

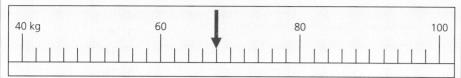

She guesses that she weighs 65 kg.

What does she weigh?

How close is her guess?

Solution

There are 10 divisions between 60 kg and 80 kg, so each division is 2 kg.

There are 4 divisions from 60 kg to the pointer: 4 divisions are 8 kg.

The weight is $60 + 8 = 68$ kg.

Her guess is 65 kg. $(68 - 65 = 3)$

This is 3 kg less than her actual weight.

You can use the same approach when looking at other scales for capacity or length.

Example 3

A mechanic needs 300 ml of oil to put in an engine.
He pours some oil from this measuring bottle.

How much oil will be left after he has put 300 ml in the engine?

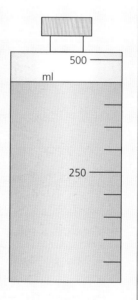

Solution

The level is between 250 ml and 500 ml.

There are 5 divisions between 250 ml and 500 ml.

Those 5 divisions cover 250 ml.

Each division is 50 ml.

The level is 4 divisions above 250 ml.

$4 \times 50 = 200$

$250 + 200 = 450$

There are 450 ml of oil in the bottle.

The mechanic pours out 300 ml.

$450 - 300 = 150$. So there will be 150 ml left in the bottle.

Example 4

This thermometer shows a child's body temperature.

The child has a fever if the temperature is 37.5 °C or higher.

What is the temperature shown on the thermometer?

Does the child have a fever?

Solution

The temperature is between 38 °C and 39 °C.

There are 10 divisions covering one degree.

Each small division is 0.1 °C.

The temperature shown is 38.2 °C.

The child does have a fever.

Example 5

Approximately how many litres are in this car's gas tank?

Solution

The tank holds 60 litres.

The tank is full when the pointer is at F, so it is between $\frac{3}{4}$ full and full.

$\frac{3}{4}$ tank is $\frac{3}{4}$ of 60 litres, which is 45 litres and full is 60 litres.

So, there is between 50 and 55 litres of fuel in the tank.

Tasks

1 Andrew has been on holiday in the UK and wants to take a bus to Heathrow Airport. He uses this timetable.

He wants to go from Exeter to Heathrow Airport.

He wants to arrive by 1.30 p.m.

Torquay		...	...	0520	0700	...	0940	1140	1400
Newton Abbot		...	...	0535	...	...	1000	1200	1420
Exeter	0140	0435	0615	0830	0945	1045	1245	1500
Heathrow Airport	0530	0805	0950	1150	1305	1410	1610	1850
London Victoria	0615	0920	1050	1250	1405	1510	1710	1950

  a What is the latest time he can catch a bus from Exeter?

  b How long will the bus journey take?

2 How much liquid is there in this bottle?

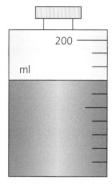

3 Kieran uses a liquid grass food on a lawn.

He pours 600 ml of liquid from this container.

How much liquid will be left after he has poured out 600 ml?

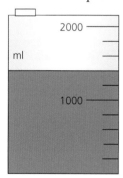

4 Ima measures her sitting room.

The pointer on this scale shows the length of one wall.

She wants to fit a sofa and a cupboard along this wall.

The sofa is 1.8 metres long and the cupboard is 1.3 metres long.

  a What is the length of the wall?

  b Is there enough space along the wall for the sofa and cupboard?

5 Yannick cuts a piece of wood to make some shelves.

The shelves will be 800 millimetres long.

The pointer on this scale shows the length of the piece of wood.

a What is the length of the piece of wood?

b How many shelves can he cut from the piece of wood?

6 This diagram shows a thermometer.

What is the temperature shown?

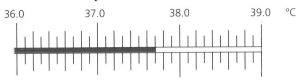

7 The table shows the population of parishes in Jamaica according to census results and official estimates.

Parish	Area (km²)	Population census 1982	Population census 1991	Population census 2001	Population census 2011	Population estimate 2017
Clarendon	1 196	203 132	214 706	237 024	245 103	247 902
Hanover	450	62 837	66 104	67 037	69 533	70 322
Kingston & St. Andrew	453	586 930	639 643	651 880	662 426	670 312
Manchester	830	144 029	159 603	185 801	189 797	192 036
Portland	814	73 656	76 317	80 205	81 744	82 710
St. Ann	1 213	137 745	149 424	166 762	172 362	174 343
St. Catherine	1 192	332 674	381 974	482 308	516 218	521 669
St. Elizabeth	1 212	136 897	145 651	146 404	150 205	151 961
St. James	595	135 959	154 198	175 127	183 811	185 846
St. Mary	611	105 969	108 780	111 466	113 615	114 959
St. Thomas	743	80 441	84 700	91 604	93 902	95 015
Trelawny	875	69 466	71 209	73 066	75 164	76 043
Westmoreland	807	120 622	128 362	138 947	144 103	145 746

What was the population of the parish of St Catherine in 2001?

8 Amber is going on holiday. She weighs her luggage.

The pointer on this scale shows the weight.

Amber can take a maximum of 23 kg of luggage.

Amber must pay $10 for every kilogram over this weight.

How much must she pay for her luggage?

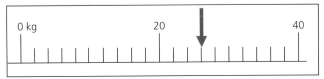

Test your knowledge

1 What is the weight shown on this scale?

a 596 g

b 560 g

c 486 g

d 480 g

2 What is the temperature on this cooking thermometer?

a 143 °C

b 146 °C

c 155 °C

d 158 °C

3 What is the amount of liquid in this measuring jug?

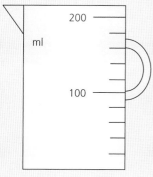

a 104 ml

b 110 ml

c 120 ml

d 140 ml

4 The table below gives the distances in kilometres (km) between some towns in northern France. How far is it from Calais to Paris?

Alençon					
365	Calais				
49	419	Le Mans			
167	433	138	Orleans		
220	293	204	135	Paris	
355	278	340	270	146	Reims
146	215	199	210	140	284	Rouen

a 135 km

b 220 km

c 293 km

d 419 km

5 This is part of a flight timetable for flights from Kingston (KIN), Jamaica to JFK Airport, New York, USA. Which is the last plane a traveller can catch to arrive in New York before 4 p.m. on a Saturday?

03:05 – 06:49	JetBlue 960	M	T	W	T	F	S	S	KIN – JFK
05:30 – 09:30	Fly Jamaica Airways 270	M	–	–	–	–	–	–	KIN – JFK
06:00 – 10:00	Fly Jamaica Airways 272	–	–	–	T	–	S	S	KIN – JFK
10:30 – 14:15	JetBlue 560	M	T	W	T	F	S	S	KIN – JFK
15:36 – 19:25	JetBlue 2960	M	–	–	T	F	S	S	KIN – JFK
19:25 – 23:10	Caribbean Airlines 17	M	T	W	T	F	S	S	KIN – JFK

a Fly Jamaica Airways 270

b Fly Jamaica Airways 272

c JetBlue 560

d JetBlue 2960

6 What is the length shown by the pointer on this scale?

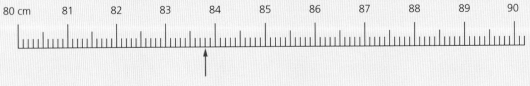

a 83.2 cm

b 83.7 cm

c 83.8 cm

d 84.3 cm

7 The diagram shows the dial on a weighing scale. What is the weight shown on the dial?

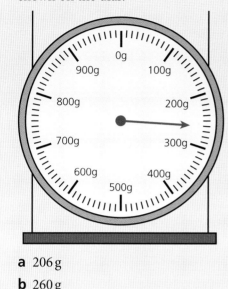

a 206 g

b 260 g

c 304 g

d 340 g

8 The table shows the top ten locations for cellphone use in the Caribbean. Which location has the highest population?

Rank	Island	Cellphone subscriptions	Population
1	Dominican Republic	4.75M	10.35M
2	Cuba	2.84M	11.05M
3	Puerto Rico	2.68M	3.62M
4	Jamaica	1.11M	2.93M
5	Haiti	1.06M	10.00M
6	Trinidad & Tobago	0.78M	1.22M
7	Guyana	0.24M	0.74M
8	Bahamas	0.23M	0.32M
9	Barbados	0.22M	0.29M
10	Suriname	0.21M	0.57M

a Dominican Republic

b Cuba

c Haiti

d Barbados

9 The diagram shows a stopwatch.

How many seconds does the stopwatch show have passed?

a 10.9

b 11.4

c 11.8

d 12.5

10 The table shows the medals won by ten countries at the athletics world championships in London, 2017.

Country	Gold medal	Silver medal	Bronze medal
1 United States	10	11	9
2 Kenya	5	2	4
3 Great Britain and NI	2	3	1
4 Poland	2	2	4
5 China	2	3	2
6 Germany	1	2	2
7 Ethiopia	2	3	0
8 France	3	0	2
9 Jamaica	1	0	3
10 South Africa	3	1	2

Which country did not win any bronze medals?

a Germany

b Ethiopia

c France

d Jamaica

Introduction

This unit is about being able to read and understand complex information. We will build on the skills you developed at Stage 2, including the work on **line graphs** in Unit 213. This unit covers extracting and interpreting information from graphs in a critical manner, identifying trends and making predictions.

In Unit 210, we introduced the common average **mean** and in this unit we will look at two other ways of expressing an average: the **mode** and the **median**. We will also look at the **range** or spread of the numbers.

You may wish to revisit Unit 210 and Unit 213 to refresh your memory about these concepts before continuing with this unit.

This unit introduces **probability** – the likelihood, possibility or chance of something happening. What is the probability of you passing your exam at the end of the course? There is a high probability, if you understand the concepts and complete the questions at the end of each unit. There is a lower probability if you do not ask for help when you do not understand.

Learning objectives

In this unit, you will find information on:

- line graphs, including trend lines

- mean, median, mode and range

- ways of expressing probability.

Line graphs

When you draw a **line graph**, make sure you label each axis clearly and include a meaningful title.

A line graph can be used to identify a trend, such as an increase in sales, and to predict a value in the future.

Multiple line graphs are drawn on the same **axes** to enable the comparison of two or more sets of data, e.g. actual sales or production figures against target sales or production figures, sales figures for the current year against the previous year, sales figures for one region against another or production at one site compared with another.

Year	India	USA	UK
2010	7.5	76	73
2012	13	83	80
2014	21	84	84
2016	30	86	89
2018		89	90

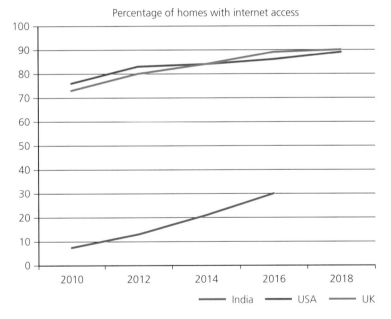

This line graph drawn from the table of data above compares the percentage of homes with internet access in India, USA and UK between 2010 and 2018. The 2018 percentage for India is not given but we can look at the trend: the percentage of homes with internet access is increasing rapidly. It is not a straight line but the percentage has more than doubled in each four-year period (2010–2014 and 2012–2016). We can extend the line and confidently predict that if this rate of growth continues, more than 40% of homes in India will have internet access in 2018.

We can also comment on the trends for UK and USA: the percentage in USA continue to increase slowly but the percentage in UK increased slightly more rapidly than USA until 2016 but appears to have reached a plateau between 2016 and 2018.

Graphs can sometimes be misleading. Look at this version of the graph.

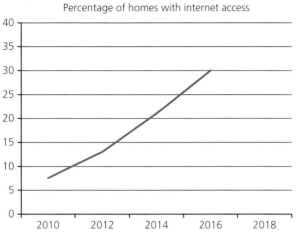

The use of a different scale in this line graph affects the slope of the line and may give the impression that the rate of increase is greater than it actually is.

Tip for assessment

Check the scale carefully when you read numbers from a line graph. For example, the vertical axis may be labelled number of people in millions meaning that when you read 5 it is actually 5 000 000.

Also, check if the scale starts at 0.

Learner tip

When you draw a graph, make sure you use a scale where you will be able to plot all your data points easily.

Mean

The **mean** average of a set of numbers can be worked out as follows:

$$\text{mean average} = \frac{\text{the sum of all the numbers}}{\text{how many numbers there are}}$$

Learner tip

You do not need to order the numbers when calculating the mean.

The total of $6 + 8 + 10 + 4 + 2$ is exactly the same as the total of $2 + 4 + 6 + 8 + 10$.

Example 1

Find the mean of 6, 8, 10, 4 and 2.

Solution

$$\text{mean} = \frac{6 + 8 + 10 + 4 + 2}{5}$$

$$= \frac{30}{5} = 6$$

So, the mean is 6

Example 2

Find the mean of 65, 84, 103, 41, 25, 58, 28, 48, 71, 27, 37, 61

Solution

First add up the total of the numbers

$65 + 84 + 103 + 41 + 25 + 58 + 28 + 48 + 71 + 27 + 37 + 61 = 648$

Then divide the total (648) by the total amount of numbers (12)

$$\frac{648}{12} = 54$$

So, the mean is 54

The answer to the mean calculation is *always* a number more than the lowest number and less than the highest number. You can check a mean calculation by reversing the calculations:

Total (648) divided by number of values (12) = mean average (54) so to check, mean average (54) times number of values (12) = total (648).

Tip for assessment

A common error made when calculating the mean is to forget to use the 0 if one is included in the numbers you are given. If, for example, the number of hours of overtime over four weeks were 8 hours, 0 hours, 7 hours and 5 hours:

divide the total (8 + 0 + 7 + 5 = 20) by 4 to find the answer of 5 hours.

If you divide by 3, you are finding the average overtime over 3 weeks.

The mean is affected by extreme values so it is not usually used in sets of data with such values because it will not give a representative average. For example, in a small business with six staff earning between $15 000 and $20 000 and one manager earning more than $100 000, the mean wage is approximately $30 000 – which is not a typical wage!

Median

The **median** of a set of data is the middle number in any given ordered data set.

To find the median, order the data from lowest to highest and the median is the number physically in the middle of the number set. Because of this, it is not skewed by extreme values (low or high) so is usually a representative average for data sets with such values.

Example 1

Find the median of 6, 8, 10, 4 and 2.

Solution

First put them in numerical order:

2   4   ⑥   8   10

Find the middle number: 6 is the median.

Example 2

Find the median of 60, 80, 100, 40, 20, 80, 30, 60, 80, 50, 80

Solution

Step 1: order the data from lowest to highest:
20, 30, 40, 50, 60, 60, 80, 80, 80, 80, 100

Step 2: Count the total number of values (11)

Step 3: Identify the middle number:
20, 30, 40, 50, 60, ⑥⓪, 80, 80, 80, 80, 100

So the median is 60.

Tip for assessment

Count the ordered values and the original values to check there are the same number in each. If not, you have made a mistake.

The median value would be a good average for an estate agent looking for the average property price in a road because it will not be skewed by one house that needs lots of repairs or one millionaire's mansion. The median might also be a better average if one person is paid a very large amount. It would not be sensible to use the median in situations where it is important to take into account all values, for example, when calculating an individual's average monthly sales figure for the purpose of bonus or commission payments.

The method shown above will only work when the data set contains an odd number of values, in that example 11 values. The method below shows you how to find the median for a data set with an even number of values.

Example 3

Find the median of this ordered data set:
20, 30, 40, 50, 60, 60, 70, 80, 80, 80, 80, 100

Solution

Step 1: Find the two central values:
20, 30, 40, 50, 60, ⑥⓪, ⑦⓪, 80, 80, 80, 80, 100

Step 2: add the two central values (60 + 70 = 130) and divide the total by 2
(130 ÷ 2 = 65)

So, the median is 65.

> You are finding the point which is halfway between the two middle numbers (which is also the mean of the two middle numbers).

Another way to find the central number(s) is to cross off one from each end of the ordered data set until you are left with either one number (for an odd set of data) or two numbers (for an even set of data): ~~20, 30, 40, 50, 60,~~ ⑥⓪, ~~80, 80, 80, 80, 100~~

Tip for assessment

In a data set with two central values, the median value will not be the same as any of the items in the data set unless both central values are the same.

Mode

The **mode** of a set of data is the value that occurs most often.

The mode is usually seen as the most straightforward of the averages to calculate, as it is simply the figure that appears most often in a data set. The mode is a representative average in large sets of data with repetitive values. It is not usually used with small sets of data, as it is more likely to give an average at the low or high end of your values, and it is not usually used in sets of data where the majority of values are unique.

For example, the mode is a good average for shoe size within a factory. Unlike the mean and median, it also guarantees an answer which is an actual value from the data set. The mode is also a common average for data which is not numerical.

Example 1

Find the mode of 8, 4, 6, 8, 1, 9, 7, 7

Solution

Order the numbers from lowest to highest (1, 4, 6, 7, 7, 8, 8, 9)

Identify the values that occur most often.

The modal values of 1, 4, 6, ⑦, ⑦, ⑧, ⑧, 9 are 7 and 8 because there are two of each in the data set.

This is known as bi-modal – having two modes.

Example 2

Find the mode of 20, 13, 18, 14, 14, 13, 20, 14, 12

Solution

Order the numbers from lowest to highest (12, 13, 13, 14, 14, 14, 18, 20, 20)

Then identify the value that occurs the most often
(12, 13, 13, ⑭, ⑭, ⑭, 18, 20, 20)

So, the mode is 14.

Tip for assessment

Be careful when you count how often each value occurs. You can identify the mode without ordering the data but this increases the chances of making a mistake. You will often have to extract the data from tables or charts and so it easy to make errors. If the data set is large, you can use a tally chart to help find the mode.

It is helpful to find a strategy to remember which method gives which average. Here is one suggestion:

Mean: the mean is mean, as in nasty, because it is the most difficult and time consuming to calculate.

Median: the median is the line which runs through the middle of a highway (in America) *or* it sounds a bit like middle.

Mode: the first two letters of mode are m and o, which stand for 'most often'.

Range

The **range** is the difference between the highest value and the lowest value in a set of numbers.

Range = highest value – lowest value

The range of a group of numbers tells us how far the values are spread. It does not tell you anything about the size of the numbers, but does tell you how far apart they are. The range of ages in a town-centre night club, for example, is likely to be small compared to the age range at a football match as football matches are watched by people of all ages.

In mathematics, the range is calculated by subtracting the smallest number from the largest number. For example, if the oldest person in a nightclub is 28 years old and the youngest is 18 then the range would be:

28 – 18 = 10 years

If the eldest supporter at the football match was 87 years old and the youngest was 3, then the range would be:

87 – 3 = 84 years

Comparing the two ranges, you can see that the first set has ages which are quite close together, but in the second set they are spread out.

Example 1

Find the range of the following car prices:

| $13 000 | $15 500 | $16 000 | $12 000 | $16 500 |

Solution

First find the highest price ($16 500)

Next find the lowest price ($12 000)

Then subtract the lowest from the highest ($16 500 – $12 000 = $4 500)

The range is $4 500

Example 2

Find the range of the following house values:

| $130 000 | $84 000 | $150 500 | $160 000 |
| $250 000 | $76 500 | $120 000 |

Solution

First find the highest house price: $250 000

Next find the lowest house price: $76 500

Then subtract the lowest from the highest $250 000 – $76 500 = $173 500)

Tip for assessment

A common error is to write the range as between 16 500–12 000, or $16 500–$12 000. Remember you need to work out the difference between the highest and lowest values.

Probability

'I may win a prize', 'there is likelihood of rain' or 'there is a good chance that football will be on the television', are all familiar phrases.

When people gamble they are competing with whoever takes the bet on **probability**. Unfortunately, many gamblers are no good at judging the odds against the calculation of probability.

For example, consider how many people would bet on whether a baby will be a boy or a girl at odds of around $\frac{7}{4}$. (That bet would pay out $7 for a correct guess for every $4 paid as a stake.) The real probability is close to $\frac{50}{50}$ for a girl or boy, so the return should more fairly be $8.

As for a lottery, the mathematical probability of matching all six numbers is 1 in 13 983 816. That is almost 1 in 14 million, so a very small probability indeed!

Real world maths

In a workplace environment, probability is used regularly. Companies can use information about their customers over a period of time and be able to predict, for example 'Thursday will be our busiest trading day', 'soup will be our most popular starter' and 'a third of today's cars will need parts ordering'. Using the probability of events happening to make successful forecasts can be extremely useful in the business world.

Learner tip

See Unit 6, Conversions between decimal fractions, common fractions and percentages for equivalent values.

Probability can be expressed in three main mathematical forms:

- a percentage
- a fraction
- a decimal.

Some people toss a coin to decide on something. If the coin is tossed fairly there is a $\frac{50}{50}$ chance of a head or tail. This is a 50% chance.

As a fraction, we would express this as a one (head) in two (head or tail) or $\frac{1}{2}$ possibility.

A decimal expression for this would be a probability of 0.5

When calculating probability in mathematics, we use a formula which enables us to handle more complex data. This formula is:

$$\frac{\text{number of stated outcomes (the thing you are looking to happen)}}{\text{number of possible outcomes (all the things that might happen)}}$$

Example 1

What is the probability of throwing a head with a coin?

Solution

$$\frac{\text{throwing a head with a coin becomes 1 stated outcome (head)}}{2 \text{ possible outcomes (head or tail)}}$$

The probability is $\frac{1}{2}$, 50% or 0.5

Probability lines

Probabilities are often shown on probability lines using either words such as **likely** and **unlikely**, or with numbers using fractions, decimals or percentages.

Note the lines go from **0 (impossible)** to **1 (certain)** with **evens** $\left(\frac{1}{2}\right)$ being exactly in the middle.

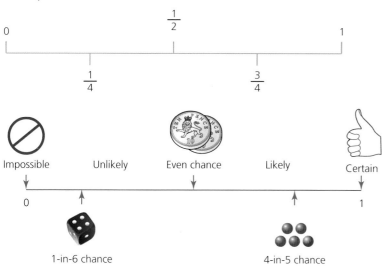

Playing cards give another familiar example that can be used for looking at probability. Successful card players often have a good memory. They remember and count the cards played and use probability to work out their next moves.

Example 2

There are 52 cards in a pack (excluding jokers). What is the probability of dealing an ace from the full pack on the first card?

Solution

Substitute a number in the formula for the stated outcome:

ace = 4 (there are 4 aces in a full pack)

Substitute a number in the formula for the possible outcomes:

full pack = 52

Then create a fraction: $\dfrac{4 \text{ stated outcomes (aces)}}{52 \text{ possible outcomes (full pack of cards)}}$

Finally, simplify the fraction: $\dfrac{4}{52} = \dfrac{1}{13}$ ←

Therefore, the probability of dealing an ace from a full pack of cards is 1 in 13. By no coincidence, 13 is also the number of cards in each playing card suit (clubs, hearts, diamonds and spades).

Note that the total of all the probabilities of things happening adds up to 1.

So in this example $\frac{5}{8} + \frac{2}{8} + \frac{1}{8}$ adds up to $\frac{8}{8}$, that is 1.

Example 3

A shoe shop takes delivery of five pairs of shoes, two pairs of boots and one pair of slippers. None of the boxes are labelled. Work out the probability that the first box opened contains:

a shoes

b boots

c slippers.

Solution

a Step 1: add the total number of all possible outcomes $(5 + 2 + 1 = 8)$

Step 2: put the number of boxes of shoes (5) into the formula

$$= \frac{5 \text{ (stated outcome)}}{8 \text{ (possible outcomes)}}$$

Step 3: express the outcome as 'a 5 in 8 chance of the first box containing shoes'.

Following the same steps for the other probabilities:

b Put the number of boxes of boots (2) into the formula

$$= \frac{2 \text{ (stated outcome)}}{8 \text{ (possible outcomes)}}$$

Simplify $\frac{2}{8}$ to $\frac{1}{4}$

Express the outcome as 'a 1 in 4 chance of the first box containing boots'.

c Put the number of boxes of slippers (1) into the formula

$$= \frac{1 \text{ (stated outcome)}}{8 \text{ (possible outcomes)}}$$

Express the outcome as 'a 1 in 8 chance of the first box containing slippers'.

The probability of something **not happening** is 1 – the probability that it does happen.

So, in this example, the probability of **not picking a box of shoes** is

$$1 - \frac{5}{8} = \frac{3}{8}$$

You can see this works as it is the probability of picking a box containing boots added to the probability of picking a box containing slippers $\left(\frac{2}{8} + \frac{1}{8} = \frac{3}{8} \right)$

Tasks

1 The number of plasters in ten different packets were as follows:

   47   47   50   46   50   49   49   47   50   47

   Find the modal number of plasters in a packet.

2 Here are the prices of the same gift in five different shops:

   $7.50 $9.00 $12.50 $10.00 $9.00

   a What is the median price?

   b In a sixth shop, the price of the gift was $11.00. What is the median price now?

3 Find the median of these numbers:

   a 1   5   2   4   8   3   1

   b 1   5   5   2   4   8   3   1

4 Find the range of these temperatures:

   $1\,°C$ $8\,°C$ $-2\,°C$ $4\,°C$ $7\,°C$ $-3\,°C$ $1\,°C$

5 Find the median and mean of this data:

   3   16   9   7   6   11   4

6 Find the mean, median and mode of this data:

   1   2   3   3   4   5

7 A gardener measures the height, in centimetres, of her sunflower plants.

   140   123   131   89   125   123   115   138

   Find the range of the heights.

8 Write down a set of five numbers with a median of 4 and a mode of 3.

9 Tami and Sean go bowling. Here are their scores:

   Tami   7   8   5   3   7

   Sean   10   8   3   1   3

   a Find the mean, median and range of each person's scores.

   b Write down two comments about their scores using your answers to part a.

   c Who would you prefer to have in your team and why?

10 Choose the best word from this scale to go with each statement below:

   Impossible   Unlikely   Evens   Likely   Certain

   a The day after Saturday will be Sunday.

   b The next baby born in Jamaica will be a girl.

   c You will wake up tomorrow morning on the moon.

11 What is the probability of having a birthday in June?

12 Complete the table below, expressing probability in different ways.

50%	$\frac{1}{2}$	1 in 2	Evens
	0.9	9 in 10	Very likely
10%	0.1	$\frac{1}{10}$	

Test your knowledge

1 Here is a set of data.

| 2 | 5 | 7 | 9 | 7 | 6 | 5 | 7 | 4 | 1 |

Find the mode of these numbers.

a 5

b 6

c 9

d 7

2 Work out the range of these numbers.

| 115 | 85 | 42 | 17 | 89 | 16 | 56 |

a 115

b 56

c 99

d 131

3 The table shows the number of TVs sold at two new superstores during their first days of trading.

| Superstore A | 12 | 12 | 7 | 3 | 13 | 4 | 13 | 10 | 6 | 14 | 12 | 7 |
| Superstore B | 4 | 5 | 7 | 20 | 3 | 9 | 7 | 5 | 5 | 13 | 2 | 6 |

Find the mode for each superstore. Work out the difference between the two modes (subtract the smaller number from the larger one).

a 5

b 7

c 6

d 13

4 An assistant records the weight, in kilograms, of people at a gym:

| 76 | 124 | 96 | 68 | 108 | 84 | 91 | 137 | 116 |

Find the mean of the weights.

a 90 kg

b 93.5 kg

c 100 kg

d 96 kg

5 Which three numbers have a mode of 64 and a mean of 72?

a 64, 72 and 80

b 64, 72 and 72

c 8, 72 and 136

d 64, 64 and 88

6 Teams A and B consist of four runners competing in a $4 \times 400\,m$ relay race. Team A takes 5 minutes and 20 seconds to complete the race and Team B takes 4 minutes and 40 seconds.

What is the difference in the mean time for a person in Team A to run 400 m and the mean time for a person in Team B to run 400 m?

a 10 seconds

b 20 seconds

c 40 seconds

d 1 minute and 20 seconds

7 Rhona carried out a survey to find out the shoe sizes of the students in her Maths class. Here is the list of everyone's shoe sizes:

3   5   7   9   7   6   5   7   4   9

The median of these numbers is:

a 5.5

b 6

c 6.5

d 7

8 The table shows the ages of a group of people:

Age	29	23	23	19	16	26	77	29	27

Find the range.

a 2

b 27

c 29

d 61

9 Express the probability 1 in 4 as a decimal.

a 0.25   **c** 0.1

b 0.4   **d** 0.14

10 If you throw a standard die, what is the probability of throwing a 6?

a 1 in 2   **c** 1 in 6

b 1 in 3   **d** 1 in 12

Unit 311
Elementary algebra

Introduction

Algebra, equations and formulae are often identified as three of the most feared topics on a mathematics curriculum. They also represent the area of mathematics that many quote as the least relevant and never used again after school. Wrong! Formulae surround our daily lives, both at work and home, and the most surprising part of this entire topic is perhaps discovering just how much knowledge of the subject you already have.

For starters, a recipe is a formula listing the ingredients you require (in the relevant proportions) to cook or bake. Flapjack = butter (200 grams) + oats (300 grams) + golden syrup (8 tablespoons). Other commonly used formulae include profit and loss, time, distance and temperature conversion. In Unit 308 and Unit 312, you will find a range of other formulae for calculating area, perimeter and volume.

Learning objectives

This unit builds on your work at Stage 2, Unit 209. At Stage 2 we solved simple equations with one unknown and substituted values into simple formulae.

In this unit you will find information on:

- more complicated formulae including those with squares, cubes and square roots
- constructing formulae and equations.

This unit looks at the relationship between distance, time and speed, which was introduced in Unit 307.

This will help you to prepare for questions about:

- drawing distance–time graphs and other graphs from given data
- using information presented in a graphical format including the idea of gradient as a rate of change.

Elementary algebra

Formulae can be expressed in two forms: written in words or summarised with letters.

For example, 'to find the perimeter of a rectangle, add the length and the width and multiply the total by two' can be simplified to: $P = 2(L + W)$

Each letter represents a word: P is perimeter, L is length and W is width.

BODMAS

We introduced **BODMAS** (or **BIDMAS**) earlier but it is very important, so here is a reminder.

BODMAS is an acronym representing:

B – Brackets

O – Orders (or I – Indices)

D – Division

M – Multiplication

A – Addition

S – Subtraction

Brackets come first in the order of operators, so any calculation given inside brackets must be completed before any other calculation. Consider the example: perimeter $= 2(L + W)$ where the length of a rectangle is 6 m and the width 3 m. Adding the two numbers in the bracket first ($6 + 3 = 9$) and then multiplying by 2, calculates the correct perimeter of 18 m. Without the brackets, however, $2L + W$ would provide a different and incorrect solution. Note the '$\times$' sign is not always necessary in a formula and multiplication should always be applied when a number is placed in front of a letter.

Orders (or Indices) are powers, which at this level only include squared and cubed. When squaring a number, multiply it by itself ($2^2 = 2 \times 2 = 4$; $3^2 = 3 \times 3 = 9$). Cubing a number requires you to multiply the number by itself and then multiply the total by itself again ($4 \times 4 \times 4 = 16 \times 4 = 64$).

The formula to find the area of a circle is area $= \pi r^2$. Using BODMAS, the r (radius) is squared before multiplying the total by π.

Note that, following through the rest of BODMAS, if presented with both $\div$ and $\times$, they are of equal status, in which case you can choose which part of the formula to do first (e.g. $6 \div 3 \times 2 = 4$). Also, $+$ and $-$ are of equal order. However, $\times$ and $\div$ are always calculated before $+$ and $-$, regardless of the order they are given.

> **Learner tip**
>
> To test a calculator, such as a version on a cellphone, try a simple calculation such as the following: $6 + 4 \div 2$ should equal 8 (divide comes before addition in BODMAS). If the answer is 5 then the answer is wrong and the calculator has added first, not following BODMAS.

Substituting values into an equation

Example 1

The selling price of an item is the total of the cost of production and the profit so the formula is:

selling price = production cost + profit

What is the selling price if the production costs are $4 and the profit is $2?

Solution

First substitute the values into the formula:

selling price = 4 + 2

So selling price = $6

Example 2

Find the perimeter (P) of a room where $L = 4\,\text{m}$ and $W = 3\,\text{m}$.

$P = 2(L + W)$

Solution

First substitute the values into the formula

$P = 2(4 + 3)$

BODMAS tells us that we need to look at the operation in brackets first.

$P = 2(7)$

When a number is next to a bracket with no operator this means multiply.

So, 2×7

Perimeter = $14\,\text{m}$

Example 3

Convert the temperature of 18 degrees Celsius into Fahrenheit.

Solution

First identify the formula: $C \times \dfrac{9}{5} + 32 = F$

Apply BODMAS: multiply 18 (C) $\times 9 = 162$

Divide $162 \div 5 = 32.4$

So $32.4 + 32 = 64.4$

$18\,°C = 64.4\,°F$

Solving simple linear equations

To solve an equation you need to rearrange the equation to have the unknown value (the symbol x, y, or whatever was used in the question) by itself on one side of the equals sign (=).

Remember, whatever you do to one side of the =, you must do the same to the other side of the =.

Example 1

$3 + x = 9$. What is the value of x?

Solution

To solve this equation you need to have x by itself on one side of the = sign.

If you take 3 away from $3 + x$ this leaves x by itself.

Then you need to do the same to the other side of the = sign. So $9 - 3$

Now you have $3 - 3 + x = 9 - 3$

$x = 9 - 3$

So $x = 6$

We can check this by putting 6 as the value of x in the original equation

$3 + 6 = 9$

This is correct, so we know the answer $x = 6$ is correct.

Example 2

$2X - 1 = 5$

What is the value of X?

Solution

To solve this equation you need to have X by itself on one side of the = sign.

If you add 1 to -1 this leaves $2X$ by itself.

Then you need to do the same to the other side of the = sign: $5 + 1$

Now you have

$2X = 5 + 1$

so $2X = 6$

Remember $2X$ is the same as $2 \times X$, and the opposite of multiplication is dividing. Divide both sides by 2 and you have $X = 6 \div 2$

$X = 3$

We can check this by putting 3 as the value of X in the original equation

$2 \times 3 - 1 = 5$. This is correct, so we know the answer $X = 3$ is correct.

Example 3

$4a - 2 = 25 + a$

What is the value of a?

Solution

To solve this equation you need to have a by itself on one side of the = sign.

If you add 2 to $4a - 2$ this leaves $4a$ by itself.

Then you need to do the same to the other side of the = sign:

$4a = 25 + a + 2$

You need all the a on the same side so take a from each side

$4a - a = 25 + a + 2 - a$

$4a - a = 3a$

$3a = 25 + 2 = 27$

$3a$ is $3 \times a$, and the opposite of multiply is divide, so we divide each side by 3:

$$\frac{3a}{3} = \frac{27}{3}$$
$$a = 9$$

We can check this by putting 9 as the value of a in the original equation

$4a - 2 = 25 + a$

$(4 \times 9) - 2 = 25 + 9$

Both sides of the = are the same, so we know the answer $a = 9$ is correct.

Construct simple formulae and equations

Consider a typical situation when calculating the rental cost of a car. There is often a standard fixed charge and then another charge which depends on the amount of gas you have used or how many days you rent the car for.

Example 1

A car company advertises a $100 fixed charge with an added daily hire rate of $50.

You plan a 3-day car rental, so you need to calculate your bill by multiplying $50 (daily rate) $\times$ 3 (number of days) and add this cost ($150) to the fixed charge ($100), to give a total overall price of $250. Create a formula to work out the total cost of the car hire.

Solution

This calculation can be more simply expressed as

overall price = fixed cost + daily rental $\times$ number of days.

We can use letters if we wish

overall price = $C + R \times D$

C = fixed costs, R = daily rental and D = number of days

When we substitute the numbers directly into the expression, this is

overall price = $100 + 50 \times 3 = \$250$

Look at the order of operations. Do you add the 100 to 50 and then multiply by 3? If you do this, the overall cost = $450, which is wrong! We know that multiplying 3 (days) by 50 (daily rate) before adding to 100 produces the real overall cost.

Congratulations – you have just used BODMAS! Note that the order of letters places M (multiplication) before A (addition).

Example 2

A salesman receives an hourly rate of pay plus a percentage of the value of the sales he makes.

a Construct a formula to calculate his gross pay (P)

b The salesman works 30 hours one week and has sales of $4000. His hourly rate is $10 and commission is 5%. What is his gross pay for the week?

Solution

a P = number of hours $\times$ hourly rate + percentage commission $\times$ sales

$P = H \times R + C \times S$

H = number of hours, R = hourly rate, C = percentage commission and S = sales

b $P = 30 \times 10 + 5\% \times 4000$

$P = 300 + 200$

$P = \$500$

Use graphical information

You are familiar with drawing graphs when you are given a range of points to plot, but what if you are only told a car travels at 40 kilometres per hour, could you draw this graph?

Example 1

a Draw a distance–time graph for a car traveling at 40 km per hour.
b Use this graph to find the distance travelled after 3 hours.
c How long does it take to travel 200 km?

Solution

a The graph will have units of time, hours, along the horizontal axis and distance up the vertical axis.
You already know two of the points. If the car travels at 40 km per hour, it will take 1 hour to travel 40 km. You also know the first point $(0, 0)$, as the car has not travelled any distance before you start. You could work out the distance travelled for 2 hours, 3 hours, etc. but it is easier to just calculate 10 hours and then join up the points you have. This is a straight-line graph as the car is travelling at the same speed for the entire journey.

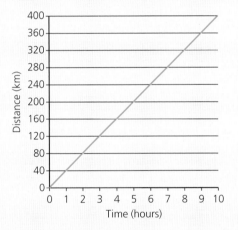

▲ Distance–time graph for a car travelling at 40 km per hour

b Draw a line up from 3 on the horizontal (or x) axis and read off the value on the vertical (or y) axis. After 3 hours the car will have travelled 120 km.

c Draw a line across from 200 on the vertical (or y) axis and read off the value on the horizontal (or x) axis. The car will have travelled 200 km after 5 hours.

Example 2

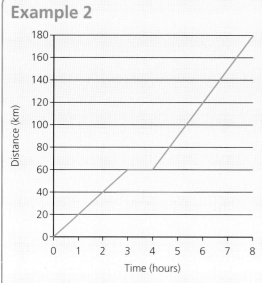

▲ Distance–time graph for a car journey

What does the graph show about the car journey?
a What speed was the car travelling for the first part of the journey?
b Why is the line horizontal at one point?
c What speed was the car travelling for the last part of the journey?

Solution

a 20 km/h
b The car was stationary.
c 30 km/h. The gradient of the line on the last part of the journey tells us that the car is travelling faster.

Activity

Construct a line graph to show 3 car journeys on the same graph.

One car travels at 30 km per hour.

The second car travels at 40 km per hour.

The third car travels at 60 km per hour.

Comment on the gradient of the lines.

Tasks

1 Calculate the volume of a cylinder with a diameter of 6 cm and a height of 20 cm.
Volume $= \pi r^2 \times h$, where h is the height of the cylinder. Assume $\pi = 3.14$

2 Mel is planning a holiday to Corfu in July. She has read that the average July temperature is 30 degrees Celsius. She prefers to know the temperature in degrees Fahrenheit and knows that the following formula will allow her to calculate this approximately:
$F = (C \times 1.8) + 32$
What will the average temperature be in °F?

3 In an electrical circuit the following formula applies:
$V = I \times R$
where V = voltage (measured in volts), I = current (measured in amps) and R = resistance (measured in ohms).
Use the formula to calculate the voltage if you have a current of 30 amps and a resistance of 8 ohms.

4 Two friends are planning a camping trip and set out to buy a tent. In order to ensure maximum comfort, they decide to calculate the volume of the tent before they buy it. They know that this formula is used to find the volume of a tent:

$$V = \frac{1}{2}(w \times h) \times L$$

If the tent measures 3 m long, 1.5 m high and 2 m wide, use the formula to calculate its volume V in m³.

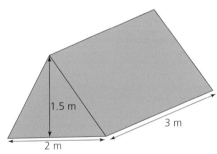

5 Solve $\dfrac{2y}{3} = 15$

6 Solve $3x - 20 = 2x$

7 Solve $4 + 2w = 10 - w$

8 What does this line graph show?

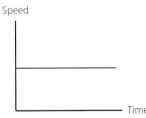

9 A delivery company charges a standard price in dollars per parcel (s) plus a fee (f) for each kilometre travelled (d).
Construct a formula to show the cost (C) of sending a parcel.

10 Solve $5x - 5 = 40$

Test your knowledge

1 Which of the following formulae is **not** correct when calculating the volume of a cuboid?

   a $V = LW + H$

   b $V = LWH$

   c $V = HWL$

   d $V = W \times L \times H$

2 If the exchange rate for pounds (£) and dollars ($) is £1 = $1.5, which of the following formulae tells you how many dollars you will get for each pound?

   a $\$ = \dfrac{£}{1.5}$

   b $£ = \$ \times 1.5$

   c $\$ = £ \times 1.5$

   d $£ = \dfrac{\$}{1.5}$

3 Your company has received an order for a circular window with a diameter of 2 m.

   How much glass you will need? Use area $= 3.14 \times r^2$

   a 12.56 m²   **c** 3.14 m²

   b 24.12 m²   **d** 6.28 m²

4 What is 20 feet in metres, approximately?

   $M = \dfrac{3f}{10}$, where M = length in metres and f = measurement in feet.

   a 3.2 m   **c** 32 m

   b 6 m   **d** 60 m

5 Solve $\dfrac{6x}{3} + 2 = 14$

   a 2   **c** 6

   b 3.5   **d** 7

6 Solve $5y - 4 = 40 + y$

   a 4   **c** 10

   b 9   **d** 11

7 Solve $2 + w^2 = 11$

   a 2   **c** 4

   b 3   **d** 5

8 You need to order bottles of orange squash for a sports event.
You devise a formula to calculate how many you need to order.
Each bottle holds 1 litre. 1 litre makes 40 individual drinks.
Use O for the number of bottles of orange squash, P for the number of people and D for the number of drinks per day.
Which of these formulae will you use to calculate how many bottles you need?

a $O = P \times D \times 40$

b $O = P \times D \div 40$

c $O = P + D \times 40$

d $O = P \times D + 40$

9 This graph shows the speed of a car. When is the car travelling the fastest?

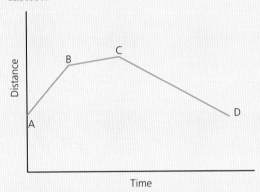

a A to B

c C to D

b B to C

d B to D

10 Which of these graphs shows a car travelling at a constant speed?

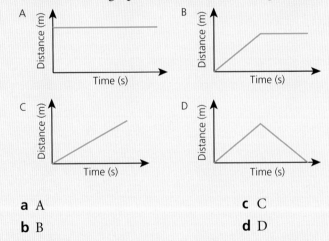

a A

c C

b B

d D

Unit 312
Shape and space

Introduction

This unit introduces the quadrilaterals:

- parallelogram
- rhombus
- trapezium

and the polygons:

- pentagon
- hexagon
- octagon.

Learning objectives

In Unit 212, we found the size of different angles by using the properties of shapes and angles on a straight line. In this unit, you will find information on:

- **alternate** and **corresponding** angles
- calculating the remaining angle in a 3-sided or 4-sided shape
- working with Pythagoras' theorem.

You will also build on work in Stage 2 to find:

- the area and perimeter of circular objects and **composite shapes**
- the volume of 3D objects with a constant cross-section.

Finally, we will use the basic ideas of similarity to compare area and volume.

You may find it useful to look through Unit 212 before you start working on this unit.

Measuring and drawing shapes

You will be familiar with a variety of shapes from Unit 104 and Unit 212. At Stage 3, you should be able to recognise and draw these shapes.

Parallelogram

A **parallelogram** is a flat four-sided shape – a quadrilateral.

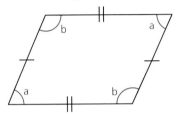

Opposite sides are parallel and equal in length (noted by the single or double short straight lines on the sides of the shape above).

Opposite angles *a* are equal and opposite angles *b* are equal.

Angles *a* and *b* add up to 180°.

Squares, rectangles and rhombuses are all parallelograms!

Rhombus

A **rhombus** is a flat four-sided shape – a quadrilateral.

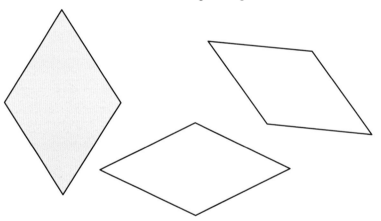

The four sides all have the same length. These shapes shown are all rhombuses. Every rhombus is a parallelogram.

A square is a quadrilateral with all sides equal in length and all interior angles are right angles. Thus a rhombus is not a square unless the angles are all right angles.

A square, however, is a rhombus since all four of its sides are of the same length.

Trapezium

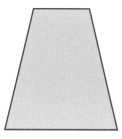 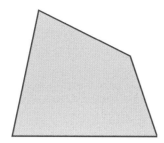

A trapezium is a flat four-sided shape – a quadrilateral.

In the Caribbean and the UK a trapezium has two parallel sides – so the shape on the left is a trapezium in the UK.

In the USA and Canada, a trapezium has no parallel sides – so the shape on the right is defined as a trapezium in the USA.

Polygons

 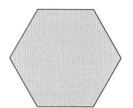 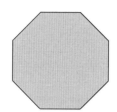

'Poly' means 'many', so a **polygon** is a shape with many sides. The common ones are:

A **pentagon** is any five-sided shape. A **regular pentagon** is one with 5 equal sides and 5 equal angles. Its interior angles are 108 degrees.

A **hexagon** is any six-sided shape. A **regular hexagon** is one with 6 equal sides and 6 equal angles. Its interior angles are 120 degrees.

An **octagon** is any eight-sided shape. A **regular octagon** is one with 8 equal sides and 8 equal angles. Its interior angles are 135 degrees.

> **Activity**
>
> Check that you know the difference between a rhombus and a square and the difference between a hexagon and a pentagon.
>
> Can you draw all the shapes?

Finding angles

You know that a square and a rectangle have four right-angles or corners. A **right-angle measures 90°.**

There are 360° in a circle and 360° in a square (4 right angles, each 90°).

We use a protractor to measure angles.

We can also work out angles from our knowledge of shapes.

The angles of a four-sided shape, e.g. a square, always add up to 360°.

> Look back at Unit 212 if you are unsure how to measure an angle.

The angles of a three-sided shape, i.e. a triangle, always add up to 180°.

If the triangle has three equal angles, what size would they each be?

$180 \div 3 = 60°$

An **equilateral triangle** has three equal sides and three equal angles.

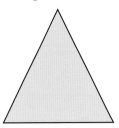

Each angle is 60°.

Not all triangles are equilateral.

Activity

Measure the angles of these shapes.

Do they add up to 360°?

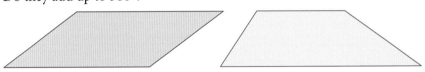

Example 1

Find the missing angle A in this triangle.

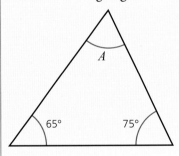

Solution

The angles of a triangle always add up to 180°.

$65 + 75 + A = 180$ or we can say $A = 180 - 65 - 75$

$A = 40°$

Example 2

Find the missing angle x in this shape.

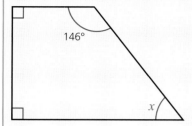

Solution

The angles of a quadrilateral always add up to 360°.

The small squares for two of the angles show these are both 90°.

$90 + 90 + 146 + x = 360$ or we can say $x = 360 - 90 - 90 - 146$

$x = 34°$

Example 3

Find the missing angle x in this shape.

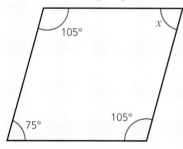

Solution

The angles of a quadrilateral always add up to 360°.

$105 + 75 + 105 + x = 360$ or we can say $x = 360 - 105 - 75 - 105$

$x = 75°$

Example 4

Find the missing angle.

Solution

Angles on a straight line add up to 180°

$180 - 140 = 40°$

So the unknown angle is 40°

Alternate and corresponding angles

In Unit 212 we looked at angles on a straight line. Here are two parallel lines.

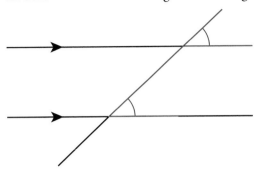

When two parallel lines are bisected by a line, the angles shown will be equal. These are known as **corresponding angles**.

We know that angles on a straight line equal 180°. In the diagram below the top angle is *a*° as shown and the angle below (*b*) is 180 − *a*. The next angle (*c*) is a corresponding angle to *a* so it is also *a*°.

This means that the angles *b* and *c* are also equal to 180°. These are known as **interior angles**.

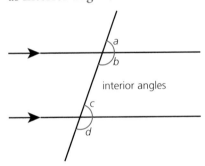

From this we can calculate the other angles.

We can use the knowledge that angles on a straight line equal 180° to work out other angles. Try it with the angles shown below.

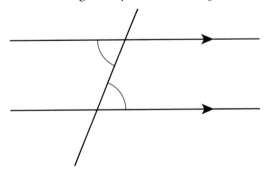

These are known as **alternate angles**.

If the statements about corresponding and alternate angles above are true, then the angles below are the same.

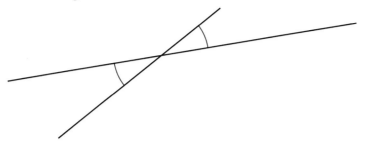

These are known as **vertically opposite angles**.

Example 1

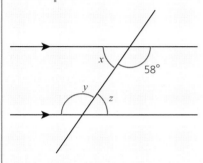

You have been given one angle, 58°.

Find angles x, y and z.

Solution

You can use your knowledge of angles to find any angle first. This is one of the solutions:

Angle x is $180° - 58° = 122°$

Angle z is a corresponding angle so $z = 58°$

Angle y is an interior angle so $y = 180° - 58° = 122°$

Activity

Draw two parallel lines and a bisecting line and check the measurements of all the angles for yourself.

Three-figure bearings

A **bearing** is an angle, measured clockwise from the North direction.

The bearing of B from A is the angle shown.

Example 1

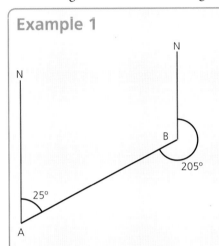

a Find the bearing of B from A.
b What is the bearing of A from B?

Solution

a To find the bearing of B from A find the degrees from North.
Use your protractor to measure 25°.
The bearing is 025° (3 figures are always given).
b The bearing of A from B is 205°.

Example 2

An aircraft leaves the airport and travels due East.

What is the bearing of the aircraft from the airport?

Solution

You can draw a diagram similar to this to help you, if you wish.

The bearing of the aircraft is 090°.

Perimeter, area and volume

In Unit 212 we calculated the perimeter of simple and composite shapes such as rectangles, squares and triangles.

At Stage 3 you need to know how to calculate the perimeter of a circle. This is called the **circumference**.

Circles

Circles are not as straightforward as the previous shapes in terms of area and perimeter. A circle can be defined as having a constant equal measurement between the boundary and the centre. The image of a circle given below is marked with the various labels required when measuring the boundary length and area.

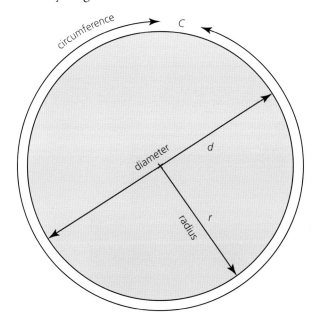

The entire measurement around the boundary of a circle is called the **circumference** (*C*).

The distance from one side to the other (measured through the centre) is the **diameter** (*d*).

The measurement taken from the centre to the boundary is known as the **radius** (*r*).

Note that the radius is exactly half of the diameter.

The following formula shows how the circumference of a circle can be calculated:

Circumference = πd

Or, because the diameter is 2 × radius,

Circumference = $2\pi r$

Learner tips

In mathematics, if there is no mathematical symbol between two terms of an equation, it always means multiply. So πd means $\pi \times$ (times) d and $2\pi r$ means $2 \times \pi \times r$.

Pi (π) is a number, denoted by a Greek letter. It is a number within a formula for calculating both the circumference and area of circles. Using the π symbol on a scientific calculator calculates more exact answers than standard calculators, as it will use the maximum number of decimal places the calculator is capable of. When using a standard calculator, you would normally round π to 3.14 (two decimal places) or 3.142 (three decimal places).

π is actually a never-ending number – the decimal places could go on forever.

Sometimes you will be told to use $\pi = \dfrac{22}{7}$ in your calculations.

Example 1

Calculate the perimeter of the following field.

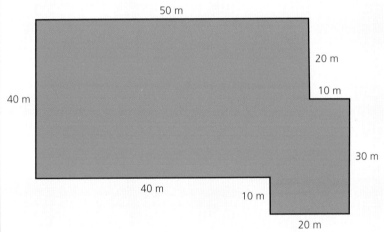

Solution

Add the lengths of each side:

$50 + 20 + 10 + 40 + 30 + 40 + 10 + 20 = 220\,\text{m}$

So, the perimeter of the field $= 220$ metres

Example 2

Find the perimeter of the triangle below measured in metres.

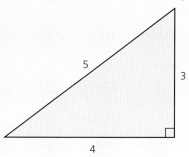

Solution

Add the lengths of each side = 5 + 4 + 3 = 12

The perimeter of the triangle is 12 metres.

Example 3

Calculate the circumference of the following wheel measured in centimetres.

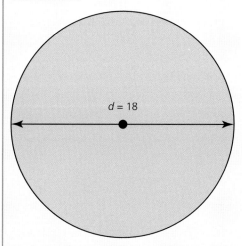

$d = 18$

Solution

To find the diameter you would draw a straight line through the centre point of the circle to the boundary either side and measure the line.

The length of the line to find the diameter is shown as 18.

The formula: circumference = πd or $2\pi r$

So substitute 3.14 (π) × 18 (diameter) or 3.14 × 2 × 9 (radius)

Calculate 3.14 × 18 = 56.52 cm

The circumference of the wheel is 56.52 cm.

Area

The **area** of a 2D shape is the space within a flat surface. This is calculated by multiplying the length and width together for a standard shape such as a rectangle.

For a rectangular room measuring 5 m (length) by 3 m (width), the area would therefore be $5 \times 3 = 15\,m^2$.

At Stage 3, calculating area is normally related to **composite shapes**, triangles or circles.

You will be expected to remember the formula for finding the area of a triangle.

Area of a triangle $= \dfrac{1}{2} \times \text{base} \times \text{height}$

Imagine a triangle as half of either a square or rectangle diagonally divided into two halves. Therefore, you can calculate the area of the rectangle and then halve the answer.

You will also need to know the formula for finding the area of a circle or a semicircle.

Area of a circle $= \pi r^2$

The area of a semicircle will be exactly half that of a circle. Therefore, when finding an area of a semicircle, treat it as an entire circle first and then divide the full area by 2.

Example 1

Find the area of the following composite shape.

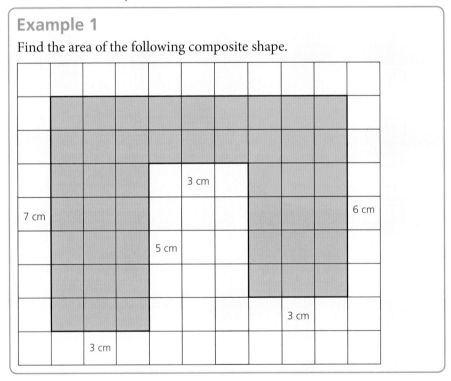

Solution

First divide the composite shape into regular shapes:

Shape A = 7 cm × 3 cm

Shape B = 3 cm × 2 cm

Shape C = 6 cm × 3 cm

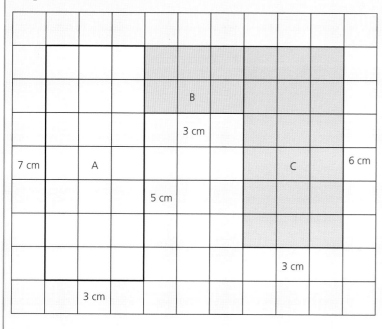

Then calculate the area of each regular shape created:

A = 7 × 3 = 21

B = 3 × 2 = 6

C = 6 × 3 = 18

Add each of three shape areas

21 (A) + 6 (B) + 18 (C) = 45

Check the units. They are all centimetres so it is cm².

The area of the composite shape is 45 cm².

Learner tips

When calculating the area of a composite shape, it can sometimes be easier to work out the area of two or more small shapes, or work out the full area of a shape and then subtract from this the area not included. Either method is acceptable.

Example 2

Calculate the area of the following triangle.

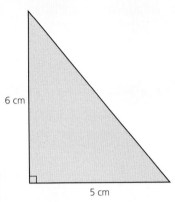

6 cm

5 cm

Solution

Identify the formula for the area of a triangle:

area $= \dfrac{1}{2} \times$ base $\times$ height

Then substitute measurements into the formula, area $= \dfrac{1}{2} \times 5 \times 6$

Multiply base and height, $5 \times 6 = 30$

Multiply 30 by $\dfrac{1}{2} = 15$ (in other words find half of 30).

The area of the triangle is 15 cm².

Example 3

Calculate the area of the following circle.

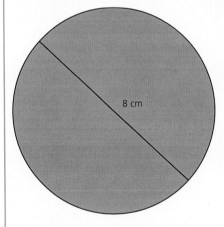

8 cm

Solution

Formula for finding the area of a circle:

area $= \pi r^2$

radius = diameter (8 cm) $\div 2 = 4$ cm

(radius can be measured from the centre point to the boundary)

Substitute 3.14 (π) and the circle radius (4 cm) so area $= 3.14 \times 4^2$

Following the BODMAS rule, square the radius first $= 4 \times 4 = 16$

Multiply 3.14 (π) by 16 (radius squared) $= 50.24$

The area of the circle is 50.24 cm².

Example 4

Find the area of the shape in the diagram below.

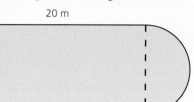

20 m

8 m

Solution

First divide the composite shape (as shown) into a rectangle and a semicircle.

Calculate the area of the rectangle,
area = 20 (length) × 8 (width) = 160

Identify the diameter of the circle = 8

Find the radius of the circle:

8 (diameter) ÷ 2 = 4 (radius)

Area of a circle: area = πr^2

Substitute 3.14 (π) and the circle radius (4 cm):

area = 3.14×4^2

Following the BODMAS rule, square the radius first = 4 × 4 = 16

Then multiply 3.14 × 16 (radius squared) = 50.24

Divide the area of a full circle by two to find the area of the actual semicircle = 25.12

Add the area of the semicircle to the area of the rectangle:

160 + 25.12 = 185.12

The units are all metres so the area unit is m²

The area of the composite shape is 185.12 m²

Learner tip

Always consider composite shapes carefully in terms of creating regular shapes. The example just completed could, for example, have had semicircles of equal size on both sides of the rectangle similar to the shape of a running track. In that scenario, the semicircles would have equalled one circle and halving the area of the circle would not have been necessary.

If there are mixed units in a question you must **always** change the units into a common unit before you start adding or multiplying. For example, if the measurement above had been 800 cm then you would convert this to 8 m and proceed as before.

Calculating the volume of 3D shapes

If we say a shape is 3D, we mean that it has three dimensions: length, width and depth.

We calculated the volume of cubes and cuboids in Unit 212. At Stage 3, calculations include the volume of a cylinder and the volume of 3D objects with a constant cross-section. In both cases, the formula is the area of the top or base × height or length.

Example 1

Calculate the volume of the following cylinder.

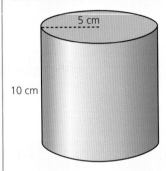

Solution

First draw a straight line through the centre point of the circle to the boundary either side. Measure the length (diameter) and divide by two to give the radius. The radius measurement of 5 cm is marked in this example.

Identify the formula for finding the volume of a cylinder:

volume = $\pi r^2 \times h$ (height)

Then substitute values into the formula

3.14 (π), 5 cm (circle radius) and 10 cm (height)

Following the BODMAS rule, square the radius first = $5 \times 5 = 25$

Multiply 3.14 (π) by 25 (radius squared) = 78.5

Multiply 78.5 (area of circular end) × 10 (height)

Check the units. As they are all centimetres, the unit of volume is cm³.

The volume of the cylinder is 785 cm³.

Cylinders are commonly used as containers for liquids, such as soft drinks. After calculating the volume of a cylinder, you can calculate its liquid content.

A volume of 1000 cm³ represents 1000 ml (1 litre of liquid).

The cylinder container in the example above would, therefore, hold 785 ml of liquid.

Tip for assessment

Always remember to indicate whether an answer is squared (2) for area or cubed (3) for volume.

Example 2

Find the volume of this metal object. The object is 10 cm long.

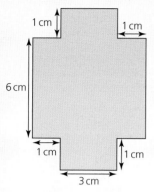

Solution

First divide the composite shape into smaller rectangles or calculate the area of the larger rectangle and then take away the area of the four corner cut-outs.

Calculate the area of the middle rectangle:

$6 \times 5 = 30 \text{ cm}^2$

Calculate the area of one of the smaller rectangles:

$3 \times 1 = 3 \text{ cm}^2$

Calculate the area of both of the smaller rectangles:

$3 \times 2 = 6 \text{ cm}^2$

Calculate total area:

$30 + 6 = 36 \text{ cm}^2$

Calculate volume:

area $\times$ height

$36 \times 10 = 360 \text{ cm}^3$

Pythagoras' theorem

Sometimes you will not be given all the sides of a triangle. Pythagoras' theorem says that, in a right-angled triangle, the area of the square of the hypotenuse is equal to the sum of the area of the squares on the other two sides. The hypotenuse is the longest side.

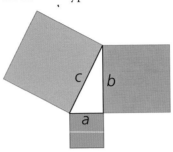

As a formula this is shown as $c^2 = a^2 + b^2$

Example 1

Find the missing side c.

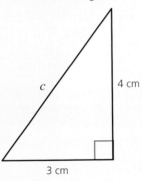

Solution

Put the numbers into the formula $a^2 + b^2 = c^2$

$3^2 + 4^2 = c^2$

Then work out the squares:

$3^2 = 3 \times 3 = 9$

$4^2 = 4 \times 4 = 16$

Add the squares:

$9 + 16 = c^2$

$25 = c^2$

Find the square root of each side:

$c = 5$ cm

Example 2

You can use the same formula to find the length of one of the smaller sides.

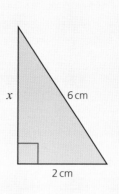

Solution

Put the numbers into the formula $a^2 + b^2 = c^2$

$x^2 + 2^2 = 6^2$

Then work out the squares:

$2^2 = 2 \times 2 = 4$

$6^2 = 6 \times 6 = 36$

$x^2 + 4 = 36$

Take 4 away from each side:

$x^2 + 4 - 4 = 36 - 4$

$x^2 = 32$

Find the square root of each side, rounded to 1 decimal place:

$x = 5.7$ cm

Basic ideas of similarity

In mathematics, shapes are similar if they have the same shape but are different sizes.

These squares are similar. One shape has sides two times the length of the other.

These triangles are similar. Although one is larger than the other, their angles are the same.

Example 1

Are these shapes similar?

Solution

No, they are not similar because all the sides have not increased in the same ratio. One shape is a square and the other is a rectangle.

Example 2

This cube has sides of 2 cm.

What effect does doubling the length have on:
a the area of the base?
b the volume?

Solution

a The area of the original cube is $2\,cm \times 2\,cm = 4\,cm^2$.
If we double the length of the sides, the area is $4 \times 4 = 16\,cm^2$.
Doubling the linear dimensions gives $4 \times$ surface area.

b The volume of the original cube is $2\,cm \times 2\,cm \times 2\,cm = 8\,cm^3$.
If we double the length of the sides, the volume is $4 \times 4 \times 4 = 64\,cm^3$.
Doubling the linear dimensions gives $8 \times$ volume.

Tasks

1 You need to buy a new carpet for an L-shaped room as shown below.

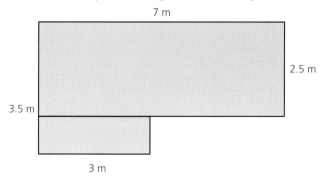

7 m

2.5 m

3.5 m

3 m

    a What is the total area of the room?

    b What is the perimeter of the room?

2 A sandpit measures $6\,\text{m} \times 250\,\text{cm}$.
   What is the area of the sandpit?

3 What is the value of the angles a, b and c in the image below?

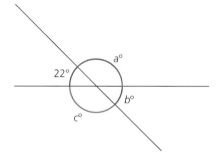

4

    a What is this shape?

    b What is the size of each angle?

5 What is the area of a circle with a diameter of $6\,\text{cm}$?

6 Find the area of this shape.

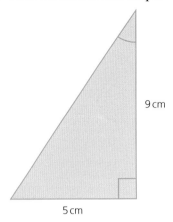

9 cm

5 cm

7 What is the volume of the shoe box?

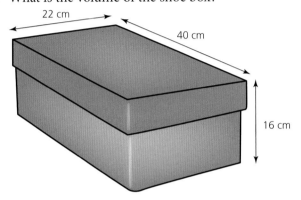

22 cm

40 cm

16 cm

8 Find the area of the shape in the diagram below.

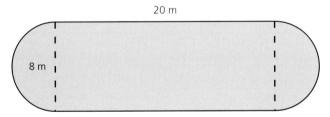

20 m

8 m

9 Is a rhombus a square? Give reasons for your answer.

10 A plane takes off from an airport and flies on a bearing of 135°.
Draw a diagram to show this.
What direction is the plane flying? (Use the points of the compass N, NE, E, SE, S, SW, W or NW.)

Test your knowledge

Assume π is 3.14 if you need to use it.

1 A fish tank has sides 15 cm × 0.6 m × 25 cm long.
What is its overall volume?

  a 22 cm³

  b 225 cm³

  c 22 500 cm³

  d 225 00 cm³

2 What is the area of the circle below (assume π = 3.14)?

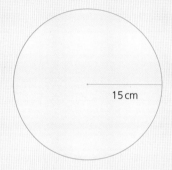

15 cm

  a 47.1 cm²

  b 94.2 cm²

  c 176.6 cm²

  d 706.5 cm²

3 A triangle has one angle of 45° and another angle of 100°.
What is the value of the third angle?

  a 35°

  b 45°

  c 55°

  d 90°

4 Find angle *x*.

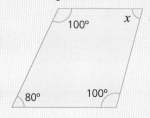

  a 60°

  b 70°

  c 80°

  d 100°

5 An aircraft leaves the airport and travels due West.
What is the bearing of the aircraft from the airport?

 a 45°

 b 90°

 c 270°

 d 315°

6 Find the missing side *x*.

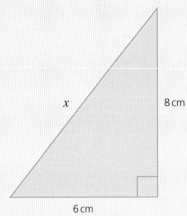

 a 8 cm

 b 9 cm

 c 10 cm

 d 11 cm

7 A cube has sides of 3 inches. If the length of the sides are doubled what effect does this have on the volume of the cube?

 a The volume is doubled

 b The volume is 3 × the original volume

 c The volume is 6 × the original volume

 d The volume is 8 × the original volume

8 A cylindrical can has a diameter of 6 cm and a height of 8 cm.
What is the volume of the can, to the nearest cubic centimetre?

   a 28 cm³

   b 48 cm³

   c 226 cm³

   d 904 cm³

9 What is the perimeter of a circular flower bed with a diameter of 12 feet?
Give your answer to the nearest foot.

   a 12 ft

   b 18 ft

   c 19 ft

   d 38 ft

10 Which of these statements is not true?

   a A parallelogram has 4 equal angles.

   b A rectangle is a parallelogram.

   c A rhombus has 4 sides of the same length.

   d The angles of a quadrilateral total 360°.

Glossary

2D having two dimensions (length and width)

3D having three dimensions (length, width and depth)

Alternate angles angles on a straight line that add up to 180°

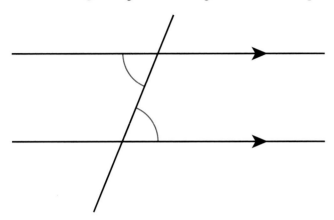

Anticlockwise the opposite direction to clockwise

Approximately fairly correct or nearly right; an estimated answer that is close to the correct answer but not exactly right

Area the space taken up by a 2D shape

Arithmetic mean average the average calculated by adding up all the number values, and dividing the total by how many numbers you have

Average the typical example of something or the representative value of a set of numbers; the mode, median or mean average can be used

Axes the lines at the side and bottom of a chart, with numbers or names; the one at the bottom is the horizontal (x) axis and the one at the side is the vertical (y) axis

Bank statement a document that shows the amount of money in your bank account, and what money has gone into and out of the account during a certain period

Bar chart a chart that presents information in a series of blocks or bars; the height of each bar represents the amount for that particular category

Bearing an angle, measured clockwise from North

BIDMAS the order of operations when carrying out a sum: Brackets, Indices, Division, Multiplication, Addition, Subtraction

Binary system the standard number system used by computers, which only uses the digits 0 and 1

Bits the digits used in the binary system (0 and 1)

BODMAS the order of operations when carrying out a sum: Brackets, Orders, Division, Multiplication, Addition, Subtraction

Capacity a measure of the amount that a container can hold

Circumference the perimeter of a circle

Clockwise the direction of a clock's hands

Composite shape a complex 2D shape that you need to break down into parts so that you can work out the area

Compound interest where interest earned is added to an account every year, so each year the interest is calculated based on the new amount

Congruent shapes shapes that are identical in shape and size

Corresponding angles where two parallel lines are bisected by a line, the angles formed above the parallel lines will be equal

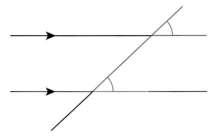

Credit money paid into a bank account

Cube number a number multiplied by itself three times, shown as ³ (for example, 2³ = 2 × 2 × 2 = 8)

Debit money taken out of a bank account

Decimal a different way of presenting fractions, using a decimal point (for example, the common fraction $\frac{1}{2}$ is 0.5 as a decimal fraction)

Decimal number system the standard number system, using the units 0–9 and then using the tens column; also called the denary system

Decimal place (d.p.) how many numbers there are on the right-hand side of the decimal point, for example 65.3 is written to 1 d.p.

Denary system the standard number system, using the units 0–9 and then using the tens column; also called the decimal number system

Denominator the bottom number in a fraction; it tells you how many parts the whole is split into

Deposit slip a form that you fill in when paying cash or cheques into a bank, also called a lodgement form

Depreciation a decrease in the value of something over time

Diameter the distance from one side of a circle to the other

Digit a figure, such as 1, 2 or 3; a number can have several digits, e.g. the number 35 has two digits

Direct proportion as one amount increases, the other amount increases at the same rate, so the ratio is always the same

Distance a measure of how far apart things are

Divisor the second fraction when you are dividing fractions (for example, in the sum $\frac{1}{2} \div \frac{1}{4}$, the divisor is $\frac{1}{4}$)

Enlargement transforming a shape by scaling it up

Equal to the same as

Equilateral triangle a triangle where all the sides are the same length and all angles are 60°

Equivalency the same value expressed in a different way, e.g. $\frac{1}{2}$ = 50% = 0.5

Estimate an approximation, not an accurate measurement

Estimating working something out roughly, not precisely

Even numbers numbers that are evenly divisible by 2 with no remainder (2, 4, 6 and so on)

Face one side of a 3D shape (including the top and bottom)

Factor a number that divides exactly into another number (for example, the factors of 6 are 1, 2, 3 and 6)

Fraction a way of describing a number that is less than one whole, for example $\frac{1}{2}$

Greater than higher or more than

Halves two equal parts of a whole; each part is one-half or $\frac{1}{2}$

Height a measure of how tall or high things are

Hexagon a 2D shape with six sides

Horizontal axis the line across the bottom of a chart

Hour hand the short hand on a clock

Hypotenuse the longest side of a right-angled triangle

Integer a whole number

Interior angles where two parallel lines are bisected by a line, the angles formed in between the parallel lines will add up to 180° (see picture)

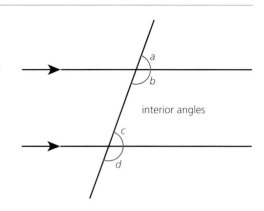

interior angles

Inverse proportion one amount decreases at the same rate that the other increases

Isosceles triangle a triangle where at least two of the sides are the same length, and at least two of the angles are the same

Legend the key to a chart, showing you what the colours mean

Length a measure of how long things are

Less than lower, or not as many

Line graph a graph that shows the line connecting a set of data points

Line of symmetry a line that you can draw through the middle of a shape so that both halves are identical

Mean average the average calculated by adding up all values in a set of data, and dividing the total by how many numbers you have in the set

Median the middle number of an ordered data set

Minute hand the pointer on a clock that shows how many minutes past the hour; it is longer than the hour hand

Mode the value that occurs most often in a set of data

Multiple a number that may be divided by another number without a remainder (for example, 18 is a multiple of 3)

Natural numbers positive whole numbers; the numbers used for counting (1, 2, 3, 4, 5 and so on)

Negative number a number below zero, shown by having a negative sign or minus sign before the number (for example −10)

Net a 2D shape that can be folded up to make a 3D shape

Numerator the top number in a fraction; it tells you how many parts the fraction shows

Odd numbers numbers that are not evenly divisible by 2 (1, 3, 5 and so on)

Parallelogram a 2D shape with 4 sides where the opposite sides are parallel and equal in length (squares and rhombuses are all parallelograms)

Pentagon a 2D shape with 5 sides

Percentage part of a whole, shown 'out of 100', for example 50% is 50 out of 100, which is the same as $\frac{1}{2}$

Perimeter the length of measurement around an entire 2D shape

Pie chart a chart that displays data as a circle divided into sectors representing the proportions of the total

Place value which digit of a number shows the number of units, which shows the number of tens and which shows the number of hundreds (and so on)

Points of the compass the directions North, East, South and West

Polygon a shape with many sides

Positive integer a whole number that is greater than zero (1, 2, 3, 4 and so on)

Prime numbers special numbers that can only be divided by themselves or by 1

Probability the likelihood, possibility or chance of something happening

Radius the distance from the middle of a circle to its side

Range the difference between the highest and lowest value in a set of numbers

Ratio a way of comparing quantities by their proportion to each other, for example if you dilute cordial with three times as much water, the ratio is 1 : 3

Reciprocal the flipped-over form of a fraction (for example, the reciprocal of $\frac{1}{4}$ is $\frac{4}{1}$)

Rectangle a 2D shape with 4 sides and 4 corners at 90° angles

Recurring decimal when the same remainder keeps recurring (for example, $1 \div 3 = 0.333$ recurring)

Reflection transforming a shape so it is like a reflection in a mirror

Regular hexagon a 2D shape with 6 equal sides and 6 equal angles

Regular octagon a 2D shape with 8 equal sides and 8 equal angles

Regular pentagon a 2D shape with 5 equal sides and 5 equal angles

Repeating decimal another name for a recurring decimal, when the same remainder keeps recurring (for example, $1 \div 3 = 0.333$ recurring)

Rhombus a 2D shape with 4 sides that are all the same length

Right angle a 90° angle, for example the angle in the corner of a square

Rotation the process or act of turning or circling around something; turning a shape round

Rounding giving a number to the nearest 10 or 100, instead of the exact number

Scientific notation a way of writing down very large or very small numbers using powers of 10 (for example $50\,000 = 5 \times 10^4$); also called standard form

Second hand a third hand on some clocks; it shows the number of seconds and is usually thinner than the hour and minute hands

Similarity in mathematics, shapes are similar if they have the same shape but are different sizes

Simple interest a percentage of the amount invested (for example, if you invested $100 in an account paying 2% interest, you would earn $2 each year)

Square a special type of rectangle where all 4 sides are the same length

Square number a number multiplied by itself or 'squared', shown as ^{2} (for example, $2^2 = 2 \times 2 = 4$)

Square root the 'root' of a square number, shown with the symbol $\sqrt{}$ (for example, $\sqrt{4} = 2$)

Standard form a way of writing down very large or very small numbers using powers of 10 (for example $50\,000 = 5 \times 10^4$); also called scientific notation

Straight line depreciation when the amount of depreciation is the same each year

Tally where you write down a stroke for each item you count, but for the fifth item counted, you 'cross' the four lines to make a gate shape

Temperature a measure of how hot things are

Tessellate fit together leaving no gaps

Three-dimensional (3D) having three dimensions (length, width and depth)

Time zones a region of the world where the same standard time is used

Translation moving a shape up, down or across but not altering it in any way

Trapezium a 2D shape with 4 sides, including 2 parallel sides

Two-dimensional (2D) having two dimensions (length and width)

Vertical axis the line up the side of a chart

Vertically opposite angles where two lines cross, the outside angles will be the same

Volume the space taken up by a 3D object

Weight a measure of how heavy things are

Width a measure of how wide things are

Index